Introduction

There are many cars with claims to fame, and many legends of motor sport, but the truth surely must be that the Bugatti Type 35 set both the rules and the runes for racing car design prior to the modern era. More Type 35s were built than any other racing car of its time. Today the Type 35 is still revered and a leading icon of the classic vintage supercar world – despite being almost 100 years old.

This is not just a vintage car, but the car that founded the Grand Prix era – a movement that is today a multi-billion dollar industry of global fame and Bugatti's Type 35 was crucial to its origins.

Ettore Bugatti's Type 35 was arguably, the world's first purpose-built, forensically designed, pure racing car – hence its sobriquet of the 'Grand Prix Bugatti'. Other cars claimed such, but few could match the sheer level of detail design and exquisite engineering of the nimble Type 35 – a car obsessive in its design delights but never at the cost of functionality. This was not a modified road car turned into a racer, and you could buy one for private and normal road use where it would perform without temper tantrums. Then there was the look of the thing – low, smaller than fashionable and sheer perfection from every angle with not a line that did not work. What on earth did rival car manufacturers think when hit with the shock of the new Type 35 in 1924?

Ferrari, Alfa Romeo, Delage, Fiat, Mercedes-Benz, Audi, Bentley, Aston Martin, these are the names of great marques that excelled as sporting cars and racing cars in the 1920s, 1930s and beyond. But it was the Bugatti Type 35 that defined a new engineering and design language and motor sport record. As such, this car is a lodestone in the development of the motor car in its first century; many might argue that it remains such as the motor car develops into its second century.

The Type 35 was Bugatti's most successful racing car and its variants enjoyed further victories. Type 35 won many a Grand Prix, the 1926 World Championship, five Tagra Florio victories, it took places at Le Mans and excelled in hill climbs, rallies and national club events all over Europe (and beyond). A Type 35 even raced in the outback of Australia at the long-forgotten races in Western Australia – the desert Grand Prix.

With its new features, exquisite engineering, and forensically detailed design, Ettore Bugatti's masterpiece sparked off a new world of focused car design amid the single-seater and two-seater categories of the 1920s.

The ingredients included a hollow front axle – through which the front leaf springs pass – a low-drag and ellipsoid body, alloy panels, ultra-thin chassis side members, special alloy wheel designs with the brakes insitu, advanced engine design with ball and roller-bearing crankshaft, and in later iterations, a supercharger on the 2.0-litre and 2.3-litre straight-eight. Throw in the dashboard mounted magneto drilled components, and engineering became art. Type 35 delivered handling and steering that set new standards for driving: Bugatti's Type 35, T35, or Tipo 35, is undoubtedly the vital monument in the history of engineering, industrial design and motor sport.

First launched in 1924 at the Lyon Grand Prix, the Type 35 drove and steered with standards of precision and response previously unknown. The strong, yet light chassis was flexible which greatly benefited the handling, the torque and power delivery ushered in a new era of performance. Yet under 400 of these hand-wrought

Top: Legend of excellence: 'EB' Ettore Bugatti stylized his badge from the beginning. His cars would become progressively more elegant, but they were always advanced in pure engineering terms.

Above: This Type 37 shows off the design that typified the Type 35 series. Note the narrow radiator, shorter nose and the suspension details.

The sheer joy of Bugatti – open motoring at speed in a road-going specification car – as it scythes through a bend in true Bugatti style.

A Type 51 caught going up Prescott hill. Note the larger radiator and longer body.

masterpieces were built from 1924 to 1931 in all its derivative variants.

Often raced in blue paint, but also subject to other schemes and numerous derivations, the Type 35 can only be described as a real icon and a consummate act of engineering genius. Today these cars are adored the world over and form a nuclei of still-racing, competitive classic racers that, although worth very large sums, are driven by enthusiasts and Bugatti Owners' Club members; the cars live and have spawned films, books, paintings and a range of resin, plastic and metal models.

Enthusiasm for the cars of Bugatti, the T35 and its variants, knows no social or financial barriers, the enthusiasm for Bugattis of old (and new) remains as strong as ever. Therefore it seems fitting that the Bugatti T35 and its story, should be the first book in the CarCraft series – publications that will reach from early motoring to the supercars of the modern era.

Bugatti in regular road use. This car has the wire wheels, lamps and wheel guards of the non-race, road specification.

Origins: The Brilliant Bugattis

Ettore Bugatti, 'Le Patron', polymath engineer and designer, a man of Italian ancestry and German/French location, took all aspects of car design to new heights of engineering precision; although the boat-tailed body shape he used had been seen on earlier Fiats and some British Sunbeam cars, Bugatti more than refined the idea, and the chassis beneath it. And no car had looked like this one. Yet Bugatti was not a formally trained engineer, and he was also highly artistic. Was he unusual in having the mind of the engineer, inventor and artist combined? Normally such traits can be in conflict – but not always and not in Ettore, nor in his son.

How did it happen that Ettore knew so much intuitively? How was it that he critically influenced the development of the motor car in the twentieth century?

The answer lay in a family from Milan headed by woodwork artist and furniture designer Carlo Bugatti. Carlo's son Ettore (1881–1947), embraced the old world of wood, and the new engineering age of cast iron, steel and metal work. The family's artistic prowess and geneology can be traced back to the 1400s – Ettore's grandfather was an architect and technician. Ettore was born of a family tradition in the arts, yet thrown into a new mechanical age – an industrial design revolution.

The family's uncle was the painter Giovanni Segantini and the Bugattis, led by Carlo the furniture designer and artist were immersed in Italian artistic affairs at the end of the nineteenth century. Milan was a centre of pre-avant-garde era thinking and of couture in clothes, bronze and marble, and in the new emerging industrial and mechanical themes.

Ettore's younger brother Rembrandt (1885–1926) had an interest in all things mechanical early on in his adult life, yet

Many say that the Type 13, latterly to include Brescia class, set the rules for Bugatti design and the consequent Type 35. Note the barrel-body with tumblehome and low-set stance.

turned to animals and sculpture as his passion. We can only speculate as to what cars or machines Rembrandt might have designed if he had not died a tragic premature death in his 30s, for Rembrandt had a passion for locomotives and would surely have embraced the new post-1900 era of the light car and more.

So it was Ettore Arco Isadoro Bugatti who became the industrial designer of cars, aircraft, railway locomotives, engines, and a range of auto-aero-mechanical conceptions. Ettore's son Gianoberto or Jean (1909–1939) inherited the family 'design' genes and in the 1930s designed car bodies of individual innovation – prior to his death in a car accident.

Inspired by his brother Rembrandt's drawing, Ettore experimented with an early tri-car (a motorized three-wheel 'car'). This device was of de Dion provenance but built by the engineers Prinetti and Stucchi. Ettore's brain was soon improving on the tri-car's design and at the age of just 17 he went to work for Prinetti and Stucchi as an apprentice. By the time he was 18, Ettore had designed a new engine and a new chassis which he soon (rather ambitiously) entered in the Paris-Bordeaux motor race. An accident forced his retirement – but he had proven a point. He raced his tri-car in many local and club events in 1898 and 1900.

But this was a three-wheeler – essentially a motorized tricycle – Bugatti wanted to build a four-wheel car and Prinetti and Stucchi would not sanction such a move. So Bugatti did so and funded the building of his first car through wealthy friends of his father; by late 1901, the then 20-year old Bugatti had built the one-off two-seater, four-cylinder engined car in Milan. Visually conventional, the car had advanced ideas in its engineering – which included the inlet and exhaust valves in the cylinder head – an unusual idea indeed for the time. The car was profiled at the 1901 International Exhibition in Milan. Bugatti, maverick and free thinker, was on his way.

The famous early carmaker, the Baron de Dietrich, seized the chance to represent Bugatti and sell his new car and persuaded (with funding) Bugatti to move north from Milan to Niederbronn in the Alsace region – the German/French territory that would be subjected to many changes of identity during the following decades of European conflict.

At this time of 1902, Alsace was German – which explains why Bugatti's later racing cars were painted white not blue or red – white was the national racing colour of Germany. Only latterly when Bugatti the Italian became 'French' by location, would the Bugatti colour become its famous blue – a version of French racing blue.

Above: Ettore even signed his engines. Exquisite design, and ego perhaps, being demonstrated.

Right: These early era cars display the developing design language of Bugatti and of note is the radiator/grille design as a precursor to the Type 35's horseshoe radiator shape.

As 1905 dawned, despite having built several small and large cars, de Dietrich abandoned car building and Bugatti needed a new home and a new partner. Enter Ettore's friend Emil Mathias the French car manufacturer from nearby Strasbourg; the pair soon produced a Mathias-Bugatti car named Hermes, which was chain-driven. This had a reversed camshaft in that it had a pull-rod system in the bottom of the engine, not a push-rod mechanism atop the engine.

The Mathis output was low-volume and within two years Bugatti was freelancing as a design engineer for other companies, by 1908 notably for the German Gasmotoren Farbrik Deutz of Cologne (who still make tractors). These were the first true tractors built in Europe. In 1908 a motorized four-wheel plough the Deutzer Pflug-Lokomotive was designed to be powered by a Bugatti 40hp B-type engine.

The Mathis cars are now cited as the early Type Bugattis – Types 6 and 7. Soon after came Types 8 and 9 as Deutz projects. Indeed, some experts ask if the first full Bugatti car design stemmed from Ettore's work at Deutz. But Mathis and Deutz cars were not 'pure' Bugatti cars made by Bugatti, owever they were the precursor to just such a vehicle. By 1909 Deutz and Bugatti had decided to go their separate ways.

A Bugatti car

Cars at this time were large and built for the wealthy – who could afford them. But Bugatti wanted to build smaller, more nimble, faster cars that could be raced, so he designed his own, secured backing from a German bank and a new era began.

Bugatti was influenced by the smaller, 1.2-litre Isotta-Fraschini cars designed by Cattaneo which competed in many races, notably the 1908 Coupe des Voiturettes at Dieppe. Bugatti drew up his own car design in 1909–1910 and indeed called this, the first true, all-Bugatti-built car design, the Type 10.

Weight – and the reduction of it – was to become a Bugatti obsession in car design long before later design greats focused on it.

In 1912 Bugatti also designed a car for the French car maker Peugeot – the 856cc Bébé. The 'Bébé was upright and functional but packed with detailed design feature. Of significance, licences to build Bugatti's cars were soon to be issued to Crossley in Great Britain, Diato in Italy, and Rabag in Germany. For those who believe Ferdinand Porsche was the first to offer his services as a designer and engineer across disciplines amid the emergent European motor industry and create cars for other people prior to creating his own, it is worth remembering that Ettore Bugatti did it first.

Bugatti's Type 10 prototype soon became a Type 13 production car and

The small and nimble Brescia at speed. The driver is somewhat exposed.

A more powerful Bugatti, one with the pear-shaped radiator and superb patina. All the elements of Bugatti thinking of the time are obvious.

A classic of Bugatti history – Ettore's Fiacre car for his sister Lidia, the Type 40 (chassis 40623) now driven by Julia de Baldanza. This was a more formal Bugatti, on the journey to Type 35's evolution in Ettore's mind.

boasted many features that refined existing design practice, or innovated new thoughts. It had a one-piece block with integral head, overhead valves, a bevel-driven overhead camshaft, unique curved tappets, an aluminium crankcase, bronze bearings, a four-speed gearbox (not three-speed), and a low sports-tourer body with a long bonnet and two lavishly upholstered seats mounted towards the rear. From this Type 10 prototype came actual, financed, car production at Molsheim, Alsace, in a new Bugatti factory – the first cars to be made – and were sold as derivations of the production series Type 10 that were labelled Type 13.

Type 13 to Brescia

Approximately 500 of this small Type 13 were built between 1910 and 1914. Initially they did not possess a defining Bugatti radiator design as would later appear, and they were dubbed 'bathtubs'. But they were quick, nimble and agile, and had a multi-valve 1,327cc, 1.3-litre overhead-camshaft engine that had torque and was eager to rev – a tuned engine such as this was very rare at the time. The gearbox had an unusual high-speed layshaft-type design in that its gear cogs were at the rear of the gearbox and the gear shaft rotated at a higher speed than the engine itself – in effect providing a pre-selection speed that reduced the effect and the time needed to speed up or slow down in the gear selection action. This gave very smooth gear changes and torque transfer. A reverse-pattern quarter-elliptic rear suspension design would soon be added to the Type 13 and remained a Bugatti design hallmark into the Type 35 years and beyond.

The steering was direct, the wheelbase short, and the whole car was tailored to be a driver's car – as such it was one of the first to be so. The Type 13 of 1913 onwards innovated Bugatti's own reversed-leaf spring design (semi-elliptic type) which would be found on all his subsequent cars.

Ettore enlarged the engine to 2.0 litres, and he even built a 5-litre powered version of the car with a long pointed tail for personal use and testing. He added an oil pump at the front of the cam-box, twin-magnetos came later.

What Type 13 lacked was style. But Bugatti worked on the body design and in 1913 invoked a distinctive ovoid egg-shaped radiator and cowl that fronted an ovoid body that ran back into a cockpit. Improved lubrication was also installed. The fuel tank or bolster tank remained externally mounted and strapped down. Never a beauty, Type 13 evolved. Yet every aspect of the car was deeply thought about – even the gearbox casing was cast with the Bugatti name and brand logo as an artistic expression, even if the body styling was minimal.

Bugatti's Type 13 set the standard for a new type of road and race car beyond its light car or voiturette origins. Type 13 was a small go-kart type of car that had point-and-shoot characteristics. However quick it was, it lacked ultimate high-speed power for endurance racing or driving. But larger engines would soon come.

When the First World War started in 1914, Bugatti moved not to Germany, but to Italy, then to France where he designed aircraft engines, notably an innovative straight-eight whose configuration was adopted by the Dusenberg Corporation in America. And did Voisin or Rolls-Royce learn anything from Bugatti's design and treatment of his engines' components? Some think they did.

Type 13 at speed – low, fast and very accurate. Driven by M. Ward.

Brescia – the precursor

After the war ended in 1918, Bugatti returned to car production and re-engined and enlarged the Type 13 into a series of cars with more appeal. The Type 13 won many races and took all four places at the 1921 Brescia Voiturette Grand Prix – hence the latter version of Type 13 becoming known as Brescia Bugatti. Type 13 had earlier won at Le Mans in 1920 and by this time Bugatti had perfected the use of the 16-valve head – something the wider car industry did not apply to mainstream models until the 1970s. Bugatti sold 40 of the race-type Full Brescia specification little short-chassis cars, tuned-up and tweaked for extra performance.

Converting the Brescia into a single-seat racer and installing an eight-cylinder engine for the Bugatti Works racing team seemed a good idea, but mechanical issues emerged and an expensive trip to the USA to compete in the Indianapolis race turned into a PR disaster for Bugatti when the cars failed and only one out the five Bugatti team cars finished.

Longer-wheelbase versions of the Type 13 were created as the Types 13, 16 and 17. The Type 20–23 series were the further developments. Types 11 and 12 were not used as model designations. From 1910–1920, Bugatti manufactured 435 of the Type 13–23 series cars – all based on the original underpinnings of the early Type 13, but continuously evolved and enlarged. Wooden wheels soon gave way to Rudge-type metal wheels.

Five chain-driven 5.0-litre cars (with at least one at 5.2 litre) were built by Bugatti just before the start of the First World War circa 1912, and although taking Type 13 design ingredients, the chassis was arguably a derivative of an earlier layout Bugatti had created at Deutz. Some evidence exists that three of these cars were built at Molseheim as late as 1914. Bugatti left the warzone of Alsace in the autumn of 1914.

A pre-1914 5.0 litre known as Black Bess, or car number 474, may be the most famous of the early Bugatti cars. Black-painted, the chain-driven type debuted the three-valve per cylinder head design. This car was delivered new to no less a figure than famed French pilot and racer Roland Garros; the car raced at Brooklands in the 1920s and remains in use today. Like other early 'Bugs', the brakes were less than brilliant. Ettore is recorded as having stated: 'I design my cars to go, not to stop!'

But the French racing style was to lean heavily on the brakes and demand quick retardation. This was an area Bugatti needed to work on.

David Scott Moncrieff, the respected veteran and vintage car observer, and admirer of Bugatti, stated in 1955, in his book *Veteran and Edwardian Cars* (Batsford, London), that Bugatti's 5.0 litre was one of the 'most outstanding efforts of motor engineering', and that this car was so far ahead of its time, so futuristic that it was 'virtually impossible to believe that it was designed in 1911'.

Such was the respect Bugatti's works imbued.

Under-bonnet four-cylinder Type 13 details. Note the oil overflow can.

Type 13 with the rear bolster fuel tank shows off its early, minimalist Bugatti design – less art and more engineering-focused.

In 1919 Bugatti designed a 3.0-litre engine as the next step from the 1.5-litre four-valve units. But the 3.0-litre international racing classification for engines was soon reduced to 2.0 litres, so he abandoned the 3.0 litre to develop a straight-eight 2.0 litre. Key to the engine's reliability was better lubrication – with oil jets and special channels within the engine.

The larger, full-bodied Types 22 and 23 emerged in various forms, including as a Brescia Modifée series. A longer wheelbase, four seats, and larger engines (2.0 litres and eight-cylinder) appeared across the Type and evolved into the Type 20 Series known as Tourers, ultimately as the eight-cylinder engined Type 30 of the 1920s. Type 30 was Bugatti's first volume production eight-cylinder car and must be seen as a stepping stone to the Type 35.

There followed from Type 13 and Brescia, a family of evolving Bugatti cars. Many were sold in Great Britain which became Bugatti's biggest export market at that time. The 'Bugs', as they became known in England, were driving and handling delights. They remain so to this day and an active series of vintage class races feature Type 13 and Brescia variants, often seen at the Bugatti Owners' Club home of Prescott, and at Brooklands, Chateau Impney, Goodwood and Shelsley Walsh.

These cars still have the traditional Bugatti-character of whining gear box and drive train, a wonderful engine sound and the alacrity of a hare being chased, as they scuttle up hills.

The next Bugatti, Type 35, was to be the true step-change, the true cementing of Bugatti's reputation and his marque's standing in the world. Beyond the Type 35, came the larger Bugattis of Types 38, 40, 43, 44, 49, 55 and beyond to the amazing Type 41 Royal, and the Types 57 and 59. Type 35 would be 'Grand Prix' but some of these cars were more 'Grand Tourer' and 'Grand Sport'. The classic Type 43 Grand Sport used the Type 35 supercharged eight-cylinder x2 monobloc engine and clearly shows some familiar details.

In Type 13 to Type 30, so started a Bugatti car 'identity' or house-style. Two thousand of the early-series Bugattis were built and provided the foundation for what would come next – the Type 35. Before that however, Ettore Bugatti took an interesting diversion on the road to greatness.

Perhaps encouraged by the somewhat unusual designs emanating from the cauldron of early car design that was Paris, Bugatti's thoughts continued apace.

Right: Even the gearbox casting had Bugatti's name cast into it. 'Le Patron' wanted you to know who did all this, it seems.

Below: Brescia in the bend displays the essential frontal view, suspension and wheel details, Note the barrel-body contours.

And Type 13 was now old and lacking a race-winning extra ingredient. Germany was emerging as the main design and engineering competitor in the emerging car industry of the 1920s – although the British were quickly catching up from a torpor when their first car and aircraft were French-inspired or even licence-built French designs. Even early British motorcycles had used French-derived engines.

At Paris, the aircraft builder Gabriel Voisin and Andre Lefebvre (latterly Citroën's chief engineer) were now building unusual, aerodynamic racing cars in the Laboratoire series; such cars had low, shovel-nosed fronts and due consideration of airflow, weight and structure as applied and as to influence performance and handling. Also at work in Paris was auto-aero engineer Pierre Cayla, and a Jacques Gerin – who would produce a stunning, elliptical design for a car at the height of the Aerodyne era.

Back at the start of the 1920s Bugatti was amid the collection of advanced and lateral design thinking and in 1923 he caused some shock by revealing his next step in racing car design. Type 13 was gone and a cigar-shaped, ellipsoid, one-piece (pontoon-style) car body devoid of wings, running boards, multiple panels, and other drag-inducing features. This was the Bugatti Tank (Tank de Tours) – or tank-bodied as it looked like a tank. It was a low-slung true aero-weapon of early aerodynamic, aerofoil or tear-drop car design that achieved high speeds for low power and was quite a shock in the world of vintage-appearance, upright cars that existed beyond Paris.

This Bugatti Tank was effectively a major step forwards in racing car design not least through its one-piece aerodynamic body shape, yet it was too much, too soon and under-developed and thus not successful. This resulted in Bugatti re-thinking his car design philosophy to a wider, more commercial mindset – yet never to be mass market, he would always produce the exquisite.

As a side tangent in design history, the British railway engineer Sir Nigel Grelsey would visit Bugatti in Paris and take inspiration from the Tank for his aerodynamically-skinned A4 Pacific Class locomotive variant with its own shovel-nosed styling. Bugatti would latterly design 1930s train stock with such aerodynamic features of his earlier thinking.

Back in 1923 Bugatti still wanted to build advanced cars, but he knew that they still had to appeal to the fashion of the era and not try and shake it up and risk failure. The answer was to create a new car that was perhaps less of an aerodynamic revolution, but more cohesive all round.

In Type 13 to Type 30 we see the process that led to the 1924 Type 35 – the defining Bugatti and its variants.

Charles Trevelyan in his Type 37. Note the twin aero-screens and wire wheels amid Type 37 spec design details. The car is almost a daily driver.

Design by Detail

Classic Bugatti trio at Prescott in 2018. The smaller scale of the Type 35 series is clearly illustrated.

So began the brilliance of Ettore Bugatti and his cars – that story which created refinements to design that then created new innovative design and a purity of engineering and appearance – 'Pur Sang' being the related pure blood term that was coined for such cars from the Type 10 prototype days. Design and function were emeshed into a driving machine that although sometimes temperamental, delivered not just a new marque, but a new type of car and driving amid a new design language.

The Type 35 was the product of one man's mind, and a car that excelled in all its parts. This was no fragile beauty, nor a fast truck. Instead it was a combination of forensically tuned parts and ideas that came together to produce the world's first true, pure-bred racing car that also happened to double as a volume-production, commercial project to deliver a usable road car for sale to the public – albeit it the very rich public.

Ettore loved racehorses and many people think the lines of the car show equine influence. The details of Ettore's design, hatched in the winter of 1923–1924, were indeed the shock of the new – the new high art of the possible, if you could think that far ahead.

Body & Chassis

Developed in 1923 off the back of earlier Bugattis, and the less-than-successful 'Tank' series, Bugatti honed every ingredient to produce a car, that although initially unsupercharged and single-carburettor induced, went on to write motoring and motor sport history. The rich and powerful, and the racing drivers of Europe, flocked to Bugatti to secure a Type 35. By early 1925, with some teething troubles sorted, larger engines and then supercharged engines were soon to be offered.

From the Type 35 came the consequent series, the Types 37, 39, 40, 51 and even ultimately the Jean Bugatti-influenced T59 racer in conceptual, if not direct model, chronology terms.

Bugatti did not copy the 1923 Fiat (F.I.A.T.) 805 racing car, but he may well have been inspired by it. Large, tall and of its era, the Fiat did however boast a lower line and a curved form, and an elegant pointed boat-tail design to its rear that hinted at beneficial aerodynamic effect. It must have set Ettore Bugatti thinking and soon he sketched a design for a leaner, more efficient car of such type. But Bugatti's design was lower, smaller, had less frontal area and a lower centre of gravity. His pre-war cars had contained such design themes too. Long before the Fiat 805 cited by some, Bugatti had worked with de Dietrich and we must wonder if the ellipsoid, boat-tailed, down-turned topline of the de Dietrich 30hp open skiff racer type of 1903 (which took part in the Paris-Madrid race driven by a female driver (Madame Du Gast)), was the true starting point for Ettore's 1924 sculpted style Type 35 body – and not the oft-cited big Fiat 805.

But there was more to his design than just modification and improvement. Bugatti thought of every aspect of the car's design and engineering and blended each feature and function into a totally new vision – beyond just a new design language.

Styling or even visual fashion had not been on Ettore's list of vital preferences in his previous cars, even if they'd had interesting lines and the addition of distinct, revised, egg-shaped radiator grille and some mascots and embellishments designed by his brother Rembrandt.

However, for the new car, style and its blending with function was vital. From above, the car's planform was a sharp elliptical shape like a racing boat's hull. From the side in profile it was lean and low and carved down and around at the rear into a pointed stern with a sharp edge to slice the air – long before Professor Kamm invented and tested a feature that did the same thing.

The scale and balance of the Type 35's lines, of its nine hand-formed, alloy panels, created the low nose, wheelbase, cabin and tail, all of which flowed into each other with a perfection of form – surely reflecting the Bugatti family's interest in sculpture. Ettore liked horses and some see equine forms in the car's body shape – the long bonnet and nose suggesting a powerful race horse's neck, say some observers. Whatever the esoterica, there is no doubt that the Type 35 was truly a defining act in car design. It had a flat bottom, an airflowed under-tray, a low frontal area and perfect compound curves, a neat little cockpit and that tail – which enclosed the rear suspension and reduced local air turbulence.

The new horseshoe grille was thin and delicate and became a brand hallmark that remains on today's VW-Group Bugattis. In developed Type 35s, the radiator shape would have to be modified.

With louvres in the bonnet, an aero-screen and with the special wired bodywork, this car's body design is beyond the forensic. Under the skin it is equally stunning in its design and manufacture.

An example of Bugatti's attention to detail is the wiring, seen wrapped around bolts along the lower sides, to secure the body panels during flex and retain any of the screw heads that are loose. Some owners create an elaborate zigzag pattern with the wiring; others use the original straight tension configuration. These wires should be wound under tension say some experts – otherwise what is the point.

The steering wheel was a walnut wood-rimmed, polished steel-spoked affair made up from four steamed, clamped and joined sections as a piece of pure art.

Type 35 (left) and Type 13 (right) show that the Type 35 was not that much larger than its ancestor, despite appearances. The evolution of Bugatti design is obvious when seen from above and behind.

The Type 35 series chassis – thin where possible and thicker where required. Note differential casting and scuttle design.

Molsheim masterpiece: This Bugatti is the 1927 Type 35A (seen in the accompanying photographs) on chassis number 4908 and was donated to the Bugatti Trust by Loren Cornelius in memory of his daughter Krista.

The essential cast alloy wheel with inset brake drum housing. Thirty-two or 24 screws on the rim according to date and specification.

Two-section scuttle/bulkhead design under the skin shows off intricate engineering and weight saving.

Unlike contemporary cars (even the Amilcar) the seating was not staggered, so the cabin was cosy to say the least. Drivers took precedence with a riding mechanic squashed in so tight he would have to put his arm out onto the rear bodywork. Drivers with large feet also had to remove their boots or wear racing shoes because the footwells were tight and the pedals very small – it was easy to press two pedals with one foot.

Despite such foibles, the car was a visual sensation and a driving delight – some have likened it to a road-going Supermarine Spitfire in its telepathic ability to respond to its pilot's commands, and in its sheer beauty of line.

Writing to his friend and supporter V. Junek in April 1924, Ettore enclosed a sketch of the Type 35's design. The production machine was not far removed from his freehand intuitive thoughts. Bugatti stated that the car was not just his finest yet, but also one of the best cars it was possible to make. No idle boast nor modest claim, but an entirely accurate observation. Ettore knew he had created something special – and had no false modesty in stating such.

The gear lever and the handbrake are externally mounted on the right side of the cockpit – remember, all early racing cars and high-class French cars were right-hand drive until the greater left-hand drive movement took hold in Europe in the 1930s.

Wheels

Although some Type 35s and 37s had traditional wire wheels fitted, the true Type 35 wheels were not just exquisite, they were design genius. Cast alloys, the eight-spoked wheels had the brake's drum mounted in a circular cavity that was integrally cast into the one piece wheel. The internal and external faces of the spokes were perfectly cast and machined. Cast alloy wheels were a Bugatti act of science – a motoring innovation not widely deployed beyond racing car use until the 1970s. Early racing-use Type 35 wheels were fitted with beaded-edge tyres and a retaining strip for the tyres inside the wheel – an early safety-type tyre-restrainer that ensured the tyre would not come off the rim. Front and rear wheels were identical.

Argument and theories rage over what inspired Ettore to the wheels' design, and some cite the works of the American engineer Harry Miller (another rare engineer-artist). Whatever the inspiration, Bugatti's shiny cast-alloy wheels were a hallmark that became an icon of design.

Detailing

Even the fascia or dashboard was form and function combined – the alloy panel being circularly polished to look good, but also to stop cracking by creating localized

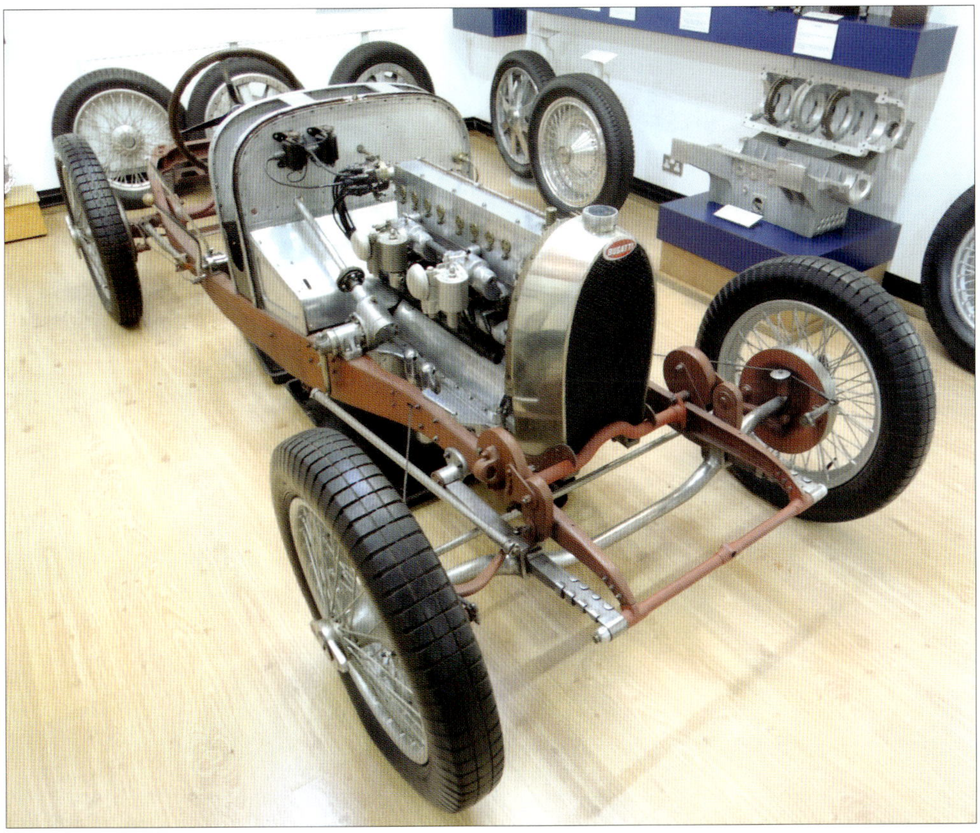

Seen at the Bugatti Trust, this is the Type 35A's delicate chassis and engine minus the elegant body. The added lightness is evident in the metal gauges and thickness ratios. The chassis longerons vary in depth and turn inwards towards the front; the engine is a structural (bolt-in) load-bearing element of the design.

crack-stopping effects devoid of corners in the metal's surface – an anti-tear effect.

The chassis beneath the Type 35's exterior was finely tuned to reduce as much weight as possible without resulting in structural fragility. As such, there was to be a degree of twist in the chassis that aided suspension dynamics during hard cornering – keeping the tyres on the tarmac, yet in extremes, allowing one wheel to lift as the opposite wheel is compressed.

Two stressed side members of mild steel in channel section formed the base of the chassis as load-carrying longerons (also seen in aircraft construction); depth varied according to need along the chassis. Under the cockpit, the channels were 170mm/6.75in section, but then tapered to near 75mm/0.75in at the front extremity of the car. Intriguingly, rather like that found in a boat (or a much later aircraft fuselage), the chassis rails changed their shape and width in planform – following the contours of the body (or hull) and being wide at mid-ships and narrower front and rear. As with the later Willys variant Jeep, a tubular cross section tied the front of the longerons together and provided bracing. At the rear, tube and channel steel added bracing and suspension location.

The engine was attached with four bolts – which under extreme crash load could snap. The engine was used as a stressed load member to assist in reducing the need for chassis gauge – as seen in 1960s Formula One racing cars and beyond.

A clever frame braced the gearbox and the whole engine-to-chassis unit provided an almost geodetic effect of stress and load

Below: Dashboard design – even the dials and gauges were designed amid a turned alloy finish.

The very thin-walled engine casting is obvious in this sectioned view of an engine. An aluminium plate closes the waterways and is screwed into position.

The eight-cylinder in-line engine block was turned and detailed in true Bugatti fashion at great expense in production costing terms. Two four-cylinder castings were set into a single or mono block configuration. Cylinders and 'head' are all-in-one.

Exquisite exhausts and very precise and expensive valve work are pure Bugatti engineering.

bearers. The scuttle had a cast bracing member at the bulkhead and was clothed in light alloy panelling.

Bugatti had designed a car in pre-war days to have a body shape integral to the chassis – long before the 1920s Fiat 805 some observers cite as of influence upon the 1924 Type 35.

Engine

Bugatti was an adherent to the single overhead camshaft and resisted changing to the twin camshaft until late 1930. So his 1,991cc straight-eight used in the Type 35 was a development of the pre-existing masterpiece seen in the Type 30 series. The aluminium crankcase – retained with the separate engine block above, was familiar. This straight-eight was mostly proven, slightly tweaked and honed to beyond exquisite under bonnet art. Bugatti's engines were straight lined, superbly dimensioned and built, highly polished, or dressed, and manufactured to forensic tolerances and measurements in expensive alloys and other metals. His use of coatings and extrusions was innovative.

The 12-part precision machined crankshaft was however all new. Two split-single rollers were introduced between the cylinders. The rollers were machined bearings and the cam profiles forensically obtained. One-piece roller bearing connecting rods of great lightness were now possible. The engine could rev out to over 5,500rpm – higher than most.

Intriguingly, the pistons were not always the same size at 26mm to 32mm, which would give differing compression ratios per piston pot.

The engine block of the Type 35 Grand Prix edition had two blocks of four-cylinders in tandem configuration. These were thin-walled integral iron head castings produced in-house at Molsheim. A top aluminium plate closes the water channels and was held down by the screwed-in vertical valve guides and the cambox. There was an internal copper oil drain and the minimal use of core plugs.

Initially unsupercharged, Bugatti's Type 35 faced stiff competition from supercharged rivals and he had to give in and go for the fitting of a compressor or supercharger – the engine shaft (or belt) driven air compressor being so much more reliable than the later turbocharger. The Paris-based, but Italian engineer Edmund Moglia did much of this work on Ettore's straight-eight engine. A Roots-type shaft driven, three-lobe compressor mounted on the right-hand side of the straight-eight was driven at engine speed via a drive shaft and two gearings. 10psi compression boosted the performance – the results of the blown application were outstanding. The extra boost was also well-balanced between the two heads' inlet process. A blower relief valve hole in the bonnet indicated that the car was supercharged.

Detail of the radiator and lamps (Marchal type) are vital elements for the Bugatti restorer and modeller to follow.

Below: A detail of the dashboard and fascia design with gearbox casting below.

Bottom: The back-to-front configured gearbox seen in exposed detail. It was a Type 30 design derivative (from earlier Brescia type design) and needed casting changes to fit the Type 35 chassis and mounting requirements. It had four speeds, not the usual three speeds.

But why did Bugatti resist the supercharger? Because it is engine-driven and not free energy like an exhaust gas-driven turbocharger. Superchargers drain horsepower out of the engine in order to drive themselves and to create forced induction. So there are losses as well as gains to the use of a supercharger. Half the gain in horespower might be lost in the mechanism. But Bugatti got the balance right.

Power was up from 90bhp/67Kw unsupercharged, to 128bhp/95Kw in the compressor or blown engine Type 35C and, 140bhp/105Kw in the ultimate 2.3-litre Type 35B. The larger 2.3-litre engines can be identified by the layman from spotting an inserted aluminium compression plate to lift the head, to allow a longer stroke, on the engine's block. Supercharged cars needed an extra fuel hand pump on the dashboard to prime the inlet manifold.

Of unusual significance are the leather couplings used in Bugatti engines – for a variety of reasons, not least to absorb loadings and flexings, they make a curious contrast with the engine's straight lines and polished parts.

The lower engine-casing lubrication was by splash effect – hence the need to warm and lubricate the engine for five minutes – especially by today's owners. The top end and camshaft were lubricated by a pump system to 15psi under centrifugal force.

Bugatti did not miss the chance to advertise on the most obvious part of the cabin – the steering wheel. More expensive design.

Above: Later cars with 330mm brake drums and extra (inverted) leaf spring (stiffener), had the leaf springs clipped at both ends, not just the front.

Top right: Vital suspension and brake design elements of the Type 35 series. Note the axles passing through the chassis. Types with the 270mm brake drums had different spring clips.

Right: The axles dips in the middle and the shock absorber is mounted directly to the chassis.

Like so many engines of its era, the straight-eight ran a bit hot and an external temperature gauge (of Boyce Moto-Meter make) was often fitted to help the driver know how hot the engine's head was.

Breathing was by Solex carburettors on the unblown Type 35A, or a Zenith (three-section fitting) to other models. Ignition was by a Scintilla magneto or latterly a Bosch system (except for the coil ignition Type 35A). The magneto was isolated from the engine bay by being placed in the dashboard. The magneto was geared, had an extra external belt drive (which could be decoupled) and operated at two speeds, dependent on the drive taken off it.

Unusually, the large fuel tank was pressurized to guarantee flow, and was a heavy steel device with internal baffles. Wooden blocks and straps were used to secure it to the chassis.

The clutch was multi-plate and extremely positive in its switch-like action. The gearbox had been proven in the Bugatti Brescia series and was a known factor of reliability. With a constant-speed, main layshaft drive, the gearbox had quick-acting characteristics, but with a reversed or upside-down gate pattern – back for First, not forwards.

Suspension

Although not a unique idea, the practice of the front axle being tubular and hollow, but then to have semi-elliptic leaf springs pass through it on forged slits (thus avoiding the need for a heavy cast U-bolt mounting), was an example of Bugatti design at its best – and patented. Special forging located the leaf springs. Expensive nickel-chrome rich steel was used. The axle was not straight and dipped at its centre to clear the chassis and rose at its ends to interact with its function at the wheel hubs. Special sliding and fixed wedges held the leaf spring and permitted castor-angle alterations (a simpler front axle was used in the Type 35A and 37).

The axle had strength in torsion and bending, yet was lightweight. The axle design reduced height and any inherent offset.

The shock absorbers were bolted to the side of the chassis frame. A small compressed C-member was internally located and loaded by a very strong tension-acting spring. There was very little rebound in the spring-to-shock absorber setting – hence the hard ride. Friction in the system seems to be its mechanism.

Good suspension behaviour requires a stiff chassis – but too stiff a chassis raises other issues; the Bugatti chassis achieves some stiffness and adds some degree of allowable compliance in twist loading. It was Ettore's genius in his location of the leaf spring and shock absorber that created the car's brilliant suspension and handling dynamics.

Effects upon the steering quality and weight from the suspension set-up were designed in, and this is why separate road and race, kingpin settings can be achieved – to alleviate the competing issues of castor angle, weight – lightness or heaviness ratio and effect.

The rear suspension was simpler – via a solid axle, tubular bracing and two radius rods that ran forwards. The leaf spring bolted to the chassis frame. Of note, a twist-resisting torque arm was fitted to the differential and ran forwards to the gearbox. Of great note, the reversed leaf springs curved or angled inwards to the stern and were hidden from external view by the bodywork. This was a very unusual feature and either follows, or permits the body design used (we shall never know which defined which). Shackles and bolts kept the springs in place yet allowed some degree of movement in the spring slot mount.

Steering

Here lies one of the car's best features – its steering, the accuracy and its feel and the ability of the system, even though it is an old-fashioned worm-and-roller wheel mechanism. The working parts are housed in an exquisite aluminium chamber and held in place by steel flanges and a leather-gloved join that is not a device to allow for misalignment, but more a way of providing a flexible element to act as an absorber between the flexural chassis, steering column reactions and the stiff, unflexing bulkhead mounting. In the supercharged

Above: The true ability of the Bugatti suspension and steering design can be seen in this shot of a turning Type 35. Think of the tyre scrub.

Left: Type 35s race to this day. Mike and Sue Marshall power up Prescott hill in their supercharged car.

variants, the steering column was shorter and the box remounted rearwards to give more engine bay room – a cast iron bracket located the column. Drop arms and links of perfect design and machining linked the steering box to its effect upon each front wheel. An exotically-curved, alloy arm was used on one side only of the system.

Of real note, the steering arms were of machined-variable size and strength – to cope with localized loadings – like the chassis itself. This was more of the mind of Ettore at work – forensic design thinking in action. The steering was direct, gave superb feedback as to what the wheels were doing and allowed the car to be placed very neatly.

Brakes

Brakes had been an issue in early sports and racing cars. Disc brakes were not in use, drum-mechanism brakes were the norm – as was cable actuation of the brakes. Reliable hydraulic brakes with servo assistance had yet to debut in motoring and Bugatti had tested an early such system to no great effect.

Bugatti designed a series of refinements to the cable-operated braking mechanism. As normal, each wheel hub carried leading and trailing brake shoes, but these were set into alloy wheel castings – heat shrunk in and riveted as well. Airflow cooling to the brakes was enhanced by such design. Clever cables with in-built tensioners and limiters would make the brakes more effective in operation and balance.

The brakes' pivot was fixed, and a cam mechanism operated the drums – adjustment and the avoidance of excess wear was vital. The brake shoes have to be symmetrically adjusted otherwise the brake system goes out of synchronization.

So Type 35 used cable-operated brakes but they were balanced by a clever gear compensator set-up to avoid one side getting more force than the other. A bevel-gear at the pedal end of the system took up the slack in the cable system and via quadrant mechanism ensured that there was no unequal play between the cables to each side's wheel brake. As with this side-to-side balancing, a front-to-rear balancing was also achieved by a chain mechanism that took up the slack in the cable system to balance the front and rear proportional braking force at the pedal. The cable cannot actuate the brakes until the chains have equalized any slack via an S-route – over sprockets that removed inherent unequal slack. This created a more equal braking force when the brake pedal was pushed.

If there was any front-axle twist encountered, this acted beneficially upon the brake mechanism that was mounted above the axle line.

The brakes at each wheel were housed in the cast-in or integral cavities within each alloy wheel – reducing weight and dynamic effects. No Ford-type cast-iron brake drum carriers here. Larger brakes were fitted to the more powerful Type 35 variants. The cable-operated brakes gave

Above: A classic view that shows the girth of the longer-nosed, broader-radiator types.

Right: The Rudge-wire wheel-spec as seen on a Type 37 with narrow boat-tail evident.

An evocative view of a classic car that set the rules for Grand Prix racing car design for years to follow.

The classic side view with spare-wheel carried, shows how short the car is, despite appearances.

Above left: The Bugatti 'hallmark' radiator and cowl have become a design icon. Adding a water temperature gauge atop the shell was a must for racers.

Above right: Smaller, lighter, less powerful, with wire wheels, yet Type 37 had purity.

excellent feedback – far more direct than a hydraulic system – but Type 35 brakes were maintenance-intensive and required regular adjustment for normal road use, let alone on the racetrack.

The handbrake was not a extra brake as we now accept such, but a direct action on the brakes without a handbrake ratchet. This gave the driver the ability to brake the rear wheels directly via a shaft-actuated external lever beside the driver and make a rally or race-style steering effect upon the rear wheels in a bend.

Type 35 was sophisticated yet not to the point of unreliability – some of its elements were simple, others were design details of genius. There were teething issues and some problems with gearboxes and cooling, but in the main, the engine was known as free-revving and easy to keep 'on the cam'.

Many patented items were within Type 35. Forensic function never matched that found in the Type 35 and this is why today it retains its supercar status amongst a world full of hypercars and more. Type 35 was driving and design at its very best prior to the age of digital authoritarianism and computer-directed design and driving. Type 35 was a true racer – maybe even to be described as a thoroughbred of a car – a racehorse that Ettore bred.

Development & Variants

Launched in 1924, little change was made to the Type in 1925, but that would soon be a year that passed, and a wide range of specification and model-range developments from early 1926 would appear. 1926 was the year when Bugatti finally accepted the concept of the forced induction engine and devised the fitting of a supercharger – but only after he had enlarged his standard engine to 2.3 litres (100mm stroke crank), but still unsupercharged, to create the 1926 model-year Type 35 T – for Targa as in the race Targa Florio.

A test car with a supercharged, reduced-stroke, sub-1.5-litre engine proved the concept. The supercharged 2-litre Type 35C was closer to the intended effect and won the Milan Grand Prix. The addition of the Roots, three-lobe 10psi supercharger and long-stroke camshaft produced the 2.3-litre Type 35TC – better known as the 35B – in 1927. A top speed of over 130mph (205km/h) and stunning acceleration times proved that swallowing his pride had been correct for Ettore when it came to supercharging – which Fiat had tried several years earlier and which even MG would soon embrace in sub-1,000cc engines.

By 1929, even more power was needed and Bugatti made a close inspection of the very powerful Miller engine – indeed swapping three Bugatti Type 43 Grand Sport cars for two Miller engines with the company itself. The Miller engine had highly efficient combustion and superb torque – Bugatti gleaned all that he could for his twin-camshaft, two-valve-per-cylinder Type 51 variant of the 35B. This late-1930 car was perhaps the ultimate development of the Type 35 design and had numerous modifications. These included a re-sited Scintilla magneto, a lower-mounted supercharger relief valve, revised wheels, and the two-filler cap style was set. The car won the 1931 race at Tunis and, with Chiron behind the wheel, won the 1931 Monaco Grand Prix. By 1935, Auto Union and Mercedes racing cars with massive engines and huge resources directed by the new Nazi German government, would sideline the lithe Bugattis. Alfa Romeo would also briefly shine with the 2.6-litre Tipo B.

The Type 35 series radiators changed as engines changed and design evolved. Many feel that the earlier narrow radiator was more elegant, yet the later, wider type is more recognizable today and is seen on the current range of modern Bugattis.

Bugatti would attempt to build bigger racing cars in the form of the Type 45 with its 16-cylinder two-bank engine, and in the 4.0-litre Type 53 with four-wheel drive. Both these types were to prove blind-alley developments and were not pursued. Even Bugatti could get it wrong.

However, enter the supercharged monster that was the Type 54 Grand Prix. Looking like a slightly fattened Type 35, this chassis and body variation on several themes, had an eight-cylinder 4,972cc, 2-valve-per-cylinder engine and a three-speed gearbox (not four speed). With 250bhp, it first debuted at Monza in 1931, but was heavy, and the steering and the rigid-sprung, live-axled suspension had a hard job keeping up with the power delivery and the handling.

Kay Don drove a T54 at over 125mph (220km/h) on the Brooklands circuit, where it quite literally flew around the banking. Today, Bugatti restorer Tim Dutton of Ivan Dutton Ltd has restored one of only five such Type 54s ever made and the car is in use at hill climbs. It goes very fast.

The Type 35 and Type 37 lent themselves to simplification or de-rating in terms of engine size and specification. Deleting alloy wheels, fitting a smaller, normally aspirated engine with less exotic and less expensive components, and lowering the trim level (such as wire wheels) produced the Grand Prix Replicas – replica in the original, manufacturers sense, not a copy or kit car as we now use the term. Type 35 Course Imitation (C.I.) or Tecla (meaning false) was the terminology for these road-going touring cars that could be used as daily drivers yet which looked like true Grand Prix racers. This de-rated gentleman's sportster sold for 30 percent less than the full-fat Type 35 and 130 of the Type 35 C.I./Tecla were manufactured.

The Type 37 in 37 and 37A guise, with 1.5-litre engines, coil ignition and the leaner bodywork, were a return to Bugatti purism in their more minimal design and effect; such cars are still popular today. They are sportsman's cars, not poseurs' cars. They are nimble and very Bugatti. 290 were made.

Although not strictly a Type 35 derivative, Jean Bugatti's 1933 Type 59 racer does look like one – it is however a different chassis design being lower, bigger, faster and more a of performance car brute. Jean Bugatti built these cars at the factory, perhaps away from his father's eyes. Ettore was often away in Paris and known for a short temper, so the Type 59 was an experiment carried out by stealth it seems. Yet the first three Type 59s were soon raced in the Spanish Grand Prix and at Monaco in 1934.

The car came as a 2.8-litre and then as a 3.3-litre eight-cylinder masterpiece of Bugatti design and Bugatti modification. The engine with its 2-4-2 crankshaft of 88mm was to be developed to 100mm, and then with a 4-4 throw crankshaft in 3.3-litre version. This created unbalanced effects and the high revving engine with

Pointed stern, mudguards (or wings) and rare white paint. Note the elliptical front-to-rear body planform.

Only detailed study can reveal the variations of development to the Type 35 design.

its need for lubrication and cooling was deemed a touch fragile. Broken crankcases on several of these engines indicated a balance and internal force issue.

Raced by the Bugatti works team, several of the Type 59s were sold off in the mid-1930s to British owners – including Bugatti Owners' Club luminary Earl Howe.

The non-expert might think Type 59 a direct Type 35 variant but it is not. It is now suggested that nine complete Type 59s were factory-built, and engines built for two more chassis; three further incomplete spares chassis are said to have been held back as Type 50 development mules or 'specials'. A total arguably 11-15 Type 59s and part-built chassis frames (as opposed to complete cars), are cited as having been suggested. Not all became complete cars. Of Type 59s, only a handful remain.

Design work and knowledge of Type 35's wheels and brakes led to the development of the dual-wire and metal-backplate load-bearing split-wheel design of the later Type 59. Brakes and wheels were an area of forensic design for Bugatti – alongside the chassis and engine.

Some say that the Type 59 was even more beautiful than the Type 35, yet we might suggest that if the Type 35 was beautiful, the Type 59 may be more of a beast. But Type 59 was probably obsolete before it was completed.

Behind such developments and derivatives, there lay background events and then specific updates across the Type 35 family up to 1931. Not widely known was Ettore's contact with the British engineer Harry Ricardo to discuss fuel octane and combustion issues for the Type 35. Methanol, aero-grade fuel octane, carburation, piston shape, cylinder head sculpting, valve, spark and fuel spray in the engine were all assessed and re-assessed.

Famous names from high society and commerce, not just racing, ordered Type 35s: people such as Sir Robert Bird, the Junek family, Glen Kidston, Earl Howe, Lord Cholmondeley and a certain Maurice Bunau-Varilla. Even Bugatti's arch-competitor Louis Delage loved the Type 35 and drove one.

Type 35 was sophisticated yet not to the point of unreliability – some of its elements were simple, others were design details of genius. Many patented items found were within it. Forensic function never matched that found in the Type 35 and this is why today it retains its supercar status amongst a world full of hypercars and more. Type 35 was driving and design at its very best prior to the age of digital authoritarianism and computer-directed design and driving. Type 35 was a true racer – a thoroughbred.

Supercharged blown cars had Zenith carburettors with an extra pump to add pressure in the system where the carburettor sits below the supercharger via a longer inlet run. An initial supercharger testing unit saw the pressure rate (psi) increased and a larger supercharger was decided upon as the factory specification.

Developments included the road-going, customer-specification Type 35s and

Two of the world's most original, remaining Grand Prix Bugattis, seen at Prescott on Bugatti Terrace. Note the beaded edge race tyre/wheel combination on the pale blue car. The red car is the famous T35C Genie. Prescott, home of the Bugatti Owners' Club, sees Type 40, Type 13, and Type 35, show off their development of Bugatti design up to the late 1920s. Note the low and lean build of the Type 35.

Developments included touring standard seating, with twin fillers on derivatives types and reverse-engineered into Type 35s on occasion. Note external operating levers and handles.

Twin filler caps were not unique to the Type 51 and some Type 35s were retro fitted with them.

derivatives such as Type 37, which were often fitted with mudguards or wings, headlamps (Marchal type) and tail lamps, and an extra aero-screen and a proper padded seat for the passenger. A full-frame rectangular touring windscreen could also be fitted. Grand Prix Type 35s had a proper driver's seat but only a vestigial riding mechanic's perch, whereas a proper passenger seat could be fitted to non-race tourers.

Detail developments saw modifications such as the number of screws around the alloy wheels' circumference was also changed in later production (32 6mm studs to 24 studs). Manufacturing issues with these Dunlop-made safety tyres of special size saw Bugatti revert to Michelin tyres at standard 28in x 4.75in/711mm x 100mm. Later Type 35C wheels had a large brake casting aperture or 330mm over the prior 270mm – to allow for bigger brakes. Just the thing for the modeller to be aware of.

The radiator and new horseshoe-shaped grille/cowl became the Bugatti hallmark – yet the original, slim, tapered radiator and cowl/shell had to be widened to fit the new modified engine and chassis space of the Type 35T, and then enlarged again for the Type 35B – to increase cooling once the bigger engines manifested. This was known as the Miramas radiator. The four-cylinder Type 37 range however, retained the narrow radiator.

Many enthusiasts think that the original, smaller, tapered radiator shell is prettier and better integrated into the overall styling.

Visually, the single fuel cap looks good, but later cars had twin fuel caps (one to fill, the other to inspect the contents level?) and some earlier cars have been retrofitted with them as well.

Cars (T35B) with the larger 330mm brakes, more power and bigger wheels, featured an extra leaf (inverted) as a stiffener in the spring and these are clipped over at the rear, not just at the front.

In certain cars, the aluminium of the body panels was replaced by steel, but this was a rare specification and one now regarded as less than pure by some Bugatti fanatics.

Other notable developments to the cars included extra fairings being added over the open cockpit, carrying straps for bags and extra spare wheels were also added. Headlamps, tail lamps and even a map-reading lamp could all be fitted. Radiators were enlarged, brakes upgraded, fuel mixtures enhanced and carburettors modified. Suspension settings were also modified for intended road use. At the factory, chassis, bodies and engines were at times mixed and matched to create new cars from new or less-new cars. All-weather kit was often applied by the owner, and proper mudguards of varying widths were a factory option.

The Michelin tyres were also substituted and today Engelbert tyres often appear on Type 35s, as do those of other tyres manufacturers.

Mrs Gentry (Gentry Restorations) checks the plugs on the supercharged Type 35. Clean-running is vital to performance.

Bugatti Type 35 and Derivatives/ Variants 1924–1931

Type 35 2.0 litre
1924–1927.
Eight-cylinder, unsupercharged. 1,991cc 2.0 litre, three-valve-per-cylinder overhead camshaft, straight eight engine with roller bearing camshaft and curved tappets. 90hp. Bore: 60mm. Stroke: 88mm. Leaf springs front and rear, hollow front axle, solid rear axle. Cable-actuated drum brakes set in cast alloy wheels. Lightweight bodywork and drilled chassis. Mechanical and magneto system. Race-standard car also sold to buyers. First car built was chassis #4323 registered 1st July 1924.

Type 35A 2.0 litre
1926–1930.
Eight-cylinder, unsupercharged. 2.0 litre, three-valve overhead camshaft, straight eight with deleted roller bearing cam of three-bearing type. 70hp. Bore: 60mm. Stroke: 88mm. Leaf springs front and rear, hollow front axle, solid rear axle. Wire wheels. For customer road use and cited as the Tecla series – defining a cheaper iteration of the purer race Type 35, as a Course/Circuit Imitation type.

Type 35C 2.0 litre Supercharged
1926–1930.
Eight cylinder, supercharged. With Roots-Type 2.0 litre, three-valve overhead camshaft, straight eight engine with roller bearing camshaft. 90hp/67kw. Bore: 60mm. Stroke: 88mm. Leaf springs front and rear, hollow front axle, solid rear axle. Cable-actuated drum brakes set in cast allow wheels. Lightweight bodywork and drilled chassis.

Type 35T 2.3 litre
Early 1926.
2,262cc, 2.3-litre enlarged eight-cylinder engine, unsupercharged. Bore: 60mm. Stroke: 100mm.

Type 35TC/Type B 2.3 litre Supercharged
Late 1926.
2,262cc. TC was initial nomenclature for Targa Compressor (supercharger). Became the definitive supercharged T35B-series race car. B series introduced 1927. 130hp/96kw. Modified-lobe Roots supercharger driven off front of crankshaft. Higher pressure ratio. Changes to track 4ft 1in at front, 3ft 11in at rear.

Type 37 1.5 litre, four cylinder
1925–1931.
Basic specification de-rated engine with plain bearings (non-roller). Wire wheels. Bore: 69mm. Stroke: 100mm. 60hp/44kw Coil ignition and lighting system added. Crankcase mounting changes to chassis due to shorter engine. A few T37s were re-bodied in closed-cabin, coupé form by Million-Guiet.

Type 37A 1.5 litre, four cylinder supercharged
1926–1931.
Type 37, with added supercharger, retains narrow radiator/cowl design. 90hp/67kw.

Type 39 1.5 litre, eight cylinder
1925–1929.
1,493cc 1.5-litre. To meet new Grand Prix Formula One engine rules as a de-bored straight-eight. Same specification as T35 and manufactured prior to Type 37/37A.

Type 39A 1.5 litre, eight cylinder Supercharged
1926–1927.
As per T39 but with supercharger. Produced for the 1926 Formula rules.

Some with 1,495cc and bore of 52mm with stroke of 88mm.

Type 43 2.3 litre, eight cylinder Supercharged
1928.
An extended-wheelbase sports-tourer variant of T35B. Modified crankcase but engine otherwise similar. Revised chassis longerons. Large cabin with extra seats.

Type 51 2.3 litre, eight cylinder Supercharged
1931.
Revised engine block of Miller-influence with twin overhead camshafts and internal differences. Retains monobloc single cast head design so is not a separate twin-cam head. Known also as Type 51 Grand Prix. First Grand Prix car to be be powered by the Miller-influenced Bugatti engine.

Note: Further variations taken from base Type 35 thinking produced as separate further types:

Type 54
1931.
4.9-litre engine from Type 50, Type 45 chassis but with layout and body aping Type 35/51.

Type 59
1933.
Not strictly a Type 35 derivative, but visually related and with similar design thinking as a race car. Derived from Type 57 design and only built in small numbers. A Jean Bugatti design of twin-cam type.

Technical Specification Bugatti Type 35B

Cost New: £1,150. Value today £2,000,000–£5,000,000
Engine: In-line, eight-cylinder ohc, cast iron block and head. In-line monobloc 2x4 cyclidners Alloy crankcase, five ball-roller bearing crankshaft
Cubic Capacity/Displacement: 2,622cc
Bore & Stroke: 60mm x 100mm
Max Power: 129bhp/95kw
Fuel System: Zenith carburettors and extra pump
Roots-Type supercharger at 10psi
Magneto-driven electrical system
Transmission: 4-Speed manual gearbox and multi-plate cast iron/steel wet clutch. Torque link damper at differential
Brakes: 330mm di. Cable operated with in-built tensioner and balancer mechanism
Tyres: 28x4.75in (711x120mm)
Steering: Worm and helical roller system
Chassis: Aluminium sheet body of steel and alloy chassis rails, cross-braced
Suspension: Inverted lead springs. Live axles front and rear. Semi elliptics pass through hollow axle at front. Rear radius rods. Friction dampers front and rear. Clips to leaves at front
Wheels: Cast alloys with integral cast brake drum housing. Eight spokes lightweight
Petrol Tank: Rear mounted, steel, 100 litres.
Max Speed: 130mph +/-
Standing Kilometre: 27 sec
Quarter Mile: 13 sec
0–60mph/120km/h: 6.9 sec
Dimensions:
Length: 145.1in/3,685mm
Width: 52.0in/1,321 mm
Front Track: 45.0in/1,143mm
Rear Track: 47.0in/1,194mm
*Front and track settings were subject to modification and variances.

Type 54 with Bugatti restorer Tim Dutton (Ivan Dutton Ltd) in command, writhes under full-power action.

Motor Sport Legend

The Type 35 and its derivatives were phenomenally successful, winning over 1,000 races in their time. They became tagged as Grand Prix and deservedly so. But the Type 35's beginning was inauspicious due to the fact that at its debut race at the 1924 Lyons Grand Prix tyre trouble caused retirement – but it was nothing to do with the advanced wheel design, but rather a tyre manufacturing issue. Bugatti had switched from Michelin to Dunlop and further work was required.

Within six weeks, a Type 35 had won in Spain on Michelins, and had won seven Grand Prix by the end of 1926. In 1925 a new world championship rating for manufacturers (cars to 2.0 litres) as opposed to drivers, came into being. Bugatti threw money at his works team.

Formula Libre and Formula One were the core of the race calender, allied to hill climb events and road rallies; with the Type 35's superior handling soon matched to supercharged power, the competing racing cars were 'blown' away especially so when Type 35 was supercharged.

The Type 35 family took the Grand Prix World Championship in 1926 after winning 351 races and setting 47 records in two years.

At the top of its era, Type 35 averaged more than a dozen race wins per month on a global reckoning. Bugatti won the Targa Florio for five consecutive years, from 1925 through 1929, with the Type 35 family. The Bugatti works team won the European Grand Prix at Spa on 20 July 1930 under the new fuel-ratified formula. Louis Chiron won six races and three major events for Bugatti in his driver-entered Type 35C by winning the world class races at Antibes, Rome, Reims, Spain and Monza. Tazio Nuvolari in his Scuderia Nuvolari Type 35C won four races that year, but 1928 was Louis Chiron's year.

The race record of the Type 35 series from 1924 to 1931 was beyond compare and remained a record of historic proportions into the modern age.

It would be decades before Formula One teams such as Ferrari, Lotus and Williams would usurp the record of Bugatti's continuous wins. The Bugatti works team of five cars had a huge profile, but so too did some of the company's customers who went racing in the cars. As an example, René Dreyfus won the 1930 Monaco Grand Prix in his privately entered Type 35B; he beat Bugatti works team driver Louis Chiron in a Type 35C by 1.8 seconds. After the victory Dreyfus was offered a works drive with the team and remained with the factory until 1935, when he joined Tazio Nuvolari at Alfa Romeo.

Adding the supercharger for the 1926 races, then the enlarged to 2.3 litres, and by 1931 the twin-cam specification as Type 51, created a warhorse of racing car.

Lower-cost developments saw unsuper-charged straight-eight engines, a 1.5-litre de-rated four-cylinder engine in the Type 37 (then a reverse-engineered Type 37A with a small supercharger). In 1924 only 16 Type 35s were sold and delivered, but in 1925, 90 were delivered. By 1926, a total of 250 cars were delivered: peak volume construction numbers were 32 cars in June 1926.

350 racing variants were built, but before all that happened, the truth was that the early Type 35 lacked ultimate race performance. Ettore Bugatti was forced to embrace the compressor or supercharger in order to boost power and performance. So he fitted a supercharger, and then enlarged the engine's cubic capacity. Paradoxically,

The Bugatti works team five-car start line-up at the Lyon Grand Prix of 1924. Not an ideal result as tyre issues stopped the cars sweeping the board. (Photo: The Bugatti Trust)

Benoist at the 1926 San Sebastian Grand Prix. Spare wheel not attached to body. (Photo: The Bugatti Trust)

Later years – Prescott as AUL 23 hares up the hill through the Esses.

he also produced a smaller-engined 1.5 litre in the Type 37 variant with normal aspiration – only then to supercharge that car!

Type 35 went on the market at £1,136 (FFR 100,000) which was enough to buy a very large house in London at the time. From 1925 onwards, Bugatti made numerous tweaks and alterations to the base specification of the Type 35. This would lead to a series of separate model derivatives that included steel-bodied (not alloy-bodied), smaller engines, de-rated items and simpler systems. The Type 35A offered buyers the looks and image of a racing car in a sports tourer daily driver. In 1926 the Type 35A cost £675 new. This was £425 less than the full-spec racing edition of the same car which was £1,000. Type 35A had a three-bearing crankshaft.

A cheaper engine block with less power, and the use of a coil ignition (deleting the expensive magneto) and a solid front axle, were deemed acceptable, but wire wheels, affected the impression somewhat. 124 of the cheaper Type 35A were sold and 87 unsupercharged Type 35 full-specification cars were sold in the same period.

Several specials, or one-off modified Type 35 types, were created for individual customers, and for racing specification. One rare car was the Type 35 that was fitted with fully enclosing front and rear body panels and wings (or fenders) to create a bulbous 1930s touring-type style draped over the lithe form of the Type 35 body.

A variety of racing, personal and national liveries were applied as paint schemes, green being a British favourite

1953 and the Bugatti Owners' Club cavalcade at Prescott with the R. Symondson car leading GNE 801. (Photo: Bugatti Owners' Club)

and red being an alternative French choice to the blue – Bugatti blue. Eliska Junek, the heroine of Czech Type 35 hill climbing and rallying, drove a bright yellow Type 35B.

But Type 35's fame was not all about racing circuit or hill-climb success. Type 35 was also about prowess on the roads of Great Britain and Europe and even as far afield as Australia. The rich and famous flocked to purchase these cars for personal use and in combined road, and race entry ownership. Of note in 1925, Cenek Junek won two international hill-climb events in his personal Type 35. Carlo Masetti won the Rome Grand Prix in February 1925 in his Type 35 and the Bugatti works team won the Targa Floria with Bartolomeo Costanti (and riding mechanic Soderini) in the Type 35 of Automobiles Ettore Bugatti.

Alessandro Consonni won the Circuito del Pozzo-Verona in March 1926 in his new Type 35. Arthur Terdich won the 1931 Australian Grand Prix at Phillip Island on 18 March 1929 in his Type 37A supercharged 1.5 litre. William Grover-Williams won the Monaco Grand Prix of 14 April 1929 in his green Type 35B. He also won the Grand Prix at Le Mans two months later. Throughout 1929, Type 35 or its variants won at least one major international class Grand Prix race per month – often several times a month. Hans Stuber won five major hill-climb events in his Type 35C in 1929.

Sir Malcolm Campbell purchased a Type 35 road car which he modified, including cutting an access hatch in the rear bodywork to create a boot/trunk space. The Vizcaya family, who were Bugatti backers, purchased more than one Type 35 and the two Vizcaya sons raced in several Type 35s.

Women at the wheel

Women took to the Type 35 too – Eliska Junek drove hers in the Targa Florio in April 1927. Helle Nice raced one and purchased it in 1930. Janine Jenky drove a Type 35B

The famous drivers who drove these cars to success included:

Phillippe Auber	Bartolomeo Costantini	Janine Jenky	Tazio Nuvolari
Robert Benoist	Albert Divo	Cenek Junek	Lord Raglan
Edouard Bret	René Dreyfus	Eliska Junek	Pierre Rey
Pierre Blaque Belair	André Dubonnet	Carl Junker	Timothy Rose Richards
Gaspare Bona	Louis Dutilleux	Glen Kidston	Jan Ripper
Milos Bondy	Anton Esterhazy	Arthur Legat	Philippe de Rothschild
Guy Bouriat	Francois Eysermann	Marcel Lehoux	Alain Szenasy
Jack Lemon Burton	George Eyston	Mario Lepori	Hans Stuber
Antonio Caliri	Philippe Eyancelin	Raymond Leroy	Arthur Terdich
Malcolm Campbell	Ernest Friderich	Count Aymo Maggi	William Grover Williams
Louis Charavel	Leonica Garnier	Pierre Marco	Achille Varzi
Jean Chassagne	Jules Goux	Emilio Materassi	Ferdinand de Vizcaya
Louis Chiron	Laszlo Hartmann	Josef Merz	Pierre de Vizcaya
Lord Cholmondeley	Huldreich Heusser	Ferdindando Minoia	Louis Wagner
Count Caberto Conelli	Earl Howe	Heinrich J. von Morgen	Tivadar Zichy

A very rare example of twin-rear wheels fitted (for extra traction) to a Type 35 series. Also seen on the later Type 50/59B of Wimille. The Scuderia Lemon Burton was well known in British Bugatti racing. (Photo: Bugatti Owners' Club)

to victory in the Coupe de Bourgogne four hours Burgundy race in 1928. Anne Cecile Rose Itier won a series of events in her Type 37 and a Type 51. Aniela d'Elern raced a Type 35 in 1929, and latterly the actress Kay Petre drove a Type 35 in competition at Brooklands – as did Eileen Elison. More recently Julia de Baldanza became a celebrated modern-era female Type 35B (chassis #4965) owner and racing driver. She says: 'The car has everything. It is so perfect in every way, and I can see equine influences in it.'

The Bugatti Trust's Curator is Angela Hucke who comes from a well-known Bugatti family and drove her father Uwe's Bugattis in races, rallies and events all over the world. Describing these cars, she said the Type 35 was 'incomparable' and 'wonderful'.

Snapshot of key events in the Type 35's motor sport record

1924:
Five-car Bugatti works team in the 1924 Lyon Grand Prix.
September. San Sebastian Grand Prix. Type 35 of Constantini leads but mechanical issues push him to second place. Fastest lap recorded.
1925:
1st Rome Grand Prix. Driver: Carlo Masetti.
1st Targa Florio. Driver: Madonie Medio.
1st Circuit of Garda. Driver: Count Aymo Maggi.
1st Brno-Sobesice-Bron Rally. Driver: Cenek Junek.
1st Dubi Rally: Driver: Cenek Junek.
1926:
1st Circuit of Pozzo-Verona, Driver: Aleesandro Consonni.
1st Rally of Roma-Valle Guila. Driver: Count Aymo Maggi.
1st Targa Floria. Driver: Bartolomeo Constantini.
1st Tripoli Grand Prix. Driver: François Eysermann.
1st Grand Prix of the Automobile Club of France, Mirimas. Driver: Joules Goux.
1st Grand Prix of Europe. Driver: Jules Goux.
1st Grand Prix of Spain. Driver: Bartolomeo Constantini.
1st Grand Prix of the Marne, Reims. Driver: Francois Lescot (Lebaudy).
1st Grand Prix of Comminges Saint Gaudens. Driver: Louis Chiron.
1st Rally Cup Acerbo-Pescara. Driver: Luigi Spinozi.
1st Grand Prix des Voiturettes, Boulougne. Driver: George Eyston.
1st Grand Prix of Italy. Driver: Jean Sabipa (Charavel).
1st Grand Prix of Milan. Driver: Bartolomeo Constantini.
1st Grand Prix of Marseille Mirimas. Driver: P. Matusserie.
1st Circuit of Garda. Driver: Count Aymo Maggi.

In the same year Cenek Junek took four first places and Elisak Junek took one first place in hill-climb/rally events in Czechoslovakia.

In 1927, Type 35 and its variants won 16 international class Grand Prix victories and seven hill-climb/rally victories. In 1928, Type 35 and its variants won 25 international class victories and seven hill-climb/rally victories. In 1929, Type 35 and its variants won 18 international class victories including the Australian Grand Prix, and eight hill-climb/rally victories. William Grover-Williams won the inaugural (1929) Monaco Grand Prix in his green Type 35B.

In 1930 Type 30 and its variants won 14 international class victories and seven hill

An interesting development with numerous additions to base specification as seen at Prescott – home of the Bugatti Owners' Club. (Photo: Bugatti Owners' Club)

climb/rally victories. In 1931, the Bugattis took five international classes (including the Australian Grand Prix for the third time) and six hill-climb/rally victories.

After the end of its front-line international Grand Prix career, the Type 35 family of cars carved a major niche in Great Britain via the Bugatti Owners' Club and its Prescott Hill Climb. Brooklands, Shelsley Walsh and numerous British hill-climb tracks and racing circuits also saw the Bugatti cars in action.

A. N. Spottiswode set the fastest time of the day (FTD) at the first Bugatti Owners' Club meeting in 1931 – winning a gold medal for a time of 23.6 seconds on the 357-yard course at Chalfont St Giles – prior to the club being based at Prescott. George Harvey Noble took a Type 25 to the Lewes Speed trial in 1935.

Charles Brackenbury got his Type 35 completely airborne on the banking at Brooklands in 1938 when winning the Whitsun Handicap of 1938. C. Lister Clark took his Type 35C to the first Prescott meeting in 1938 and was third at the autumn event.

In the mid-1930s Ettore and Jean Bugatti and the works team were regular visitors to Britain and to Prescott, where the new Bugatti Owners' Club had made a home and a wonderful hill-climb track that is in use to this day – the club celebrated its 90th anniversary in 2019. Ettore even donated a car to the club in the 1930s, and Jean Pierre Wimille made major headlines in a Bugatti T50B/50 at Prescott.

In the 1950s and 1960s, Bugatti Owners' Club events became established and a British team of racing and hill-climbing Bugaires consisting of Peter Gaskell, Mike Hatton and David Vickers-Jones who ran a Type 35 (with straight, external exhaust) and a Type 37.

Today Type 35s and their derivatives are raced in national class events at major high speed circuits and at hill climbs and international gatherings. A huge movement exists to support, race and restore and care for these cars as they approach their tenth decade. Companies like Ivan Dutton Ltd, Gentry Restorations, Jarrot Engineering, Tula Precision Ltd, Ashton Keynes Vintage Restorations, and Crossthwaite and Gardiner, all form the vital nuclei of Bugatti-experience experts who keep these cars on the road and on the track and driven in the manner for which they were designed.

Current Type 35 owners and drivers adore their cars, but are not afraid to use them as intended. Chris Jacques, owner of a 1926 Type 35, stated at Brooklands that the Type 35 is 'amazing, almost telepathic'. Mike Marshall, owner of a Type35/B said at Prescott: 'It's so fast, so accurate'. The likes of Nick Mason, Duncan Pittaway, Richard Crosthwaite, Edmund Burgess, Julian Mazjub, and a club of competitive Bugatti drivers love these cars and drive them fast at events across the vintage motor sport calender. Hamish Moffat, Neil Corner, and many more names populated the modern-era record of Type 35 driving.

Some current Type 35 aficionados suggest that the smaller, lighter Type 35s – unblown but with the bigger 2.3-litre engine – are the purest and best-handling of the bunch. Others cite the blown Type 37A. This is the same sort of argument that suggests that the early Supermarine Spitfire is nicer to fly and more agile than a later, heavier, larger Spitfire Mark V.

This is Prescott in 1976 at the height of 1970s Bugatti competitions. A French car speeds up the hill with the driver in classic Type 35 pose. (Photo: Bugatti Owners' Club)

'Genie'

Great Type 35s still race and are still driven as tourers. One noted car that has significant originality is the Type 35C registered as GNE 801 and known as 'Genie'. This was the car whose sale provided funds for the establishment of the world's leading repository of Bugatti artefacts, learning and knowledge transfer that is the Bugatti Trust, based at Prescott alongside the Bugatti Owners' Club.

'Genie' – now wearing the correct red paint – is original, unmolested and has not been denuded of its patina or its story. Thankfully, it has not been made 'perfect'. This was a Bugatti works car of 1,991cc supercharged, and the spare team car, chassis #4423 (originally 4899 but over-stamped by the factory in one of Bugatti's typical adjustments), which still holds the hill-climb record at Prescott and is still used on public roads despite being worth millions. It competed in a Grand Prix class race in Italy, but was then sold to a private owner (Count Trossi) who entered it in late 1920s hill climbs. In private ownership it then gravitated to British Bugatti racer and team director Jack Lemon Burton of Scuderia Lemon Burton. He sold the car on to an Alan Kershaw Haworth who kept it for decades and put many thousands of trouble-free miles on the car.

GNE 801 was then sold to John Bentley and is now owned by Peter Rae who has ensured its safe keeping and as near original condition as essential maintenance allows amid its utterly authentic originality. GNE's current value must be heading towards £5 million.

Stephen Gentry, the Bugatti restorer (Gentry Restorations), in the B Williams Type 35, howls up the hill in true Bugatti fashion – in the accurate way of the Type 35.

Type 51 enters Ettore's bend at Prescott in the twenty-first century. Edmund Burgess at the wheel.

Modern motor sport: later model Type 35 leads an earlier car down the Prescott hill-climb return road.

Prescott, home of the Bugatti Owners' Club. Here Type 40, Type 13 and Type 35 show off the development of Bugatti design up to the late 1920s. Note the low and lean build of the Type 35.

Top: Mike Marshall's Type 35 (supercharged spec) displays its blue in action at Prescott.

Above: Three types of the series. Left (pale blue) is the eight-cylinder 1926 Grand Prix spec car with 'race' tyres, centre (red car) is the famous and original-condition 1927 Type 35C 'Genie' and right (dark blue) a Type 37 four-cylinder car.

Right: 1926 Grand Prix Type 35 nose details. Chassis #4809 owned by Chris Jacques and sold via Malcolm Cambell's garage to Sir G. E. Duveen. Not blue, but a darker red of original Bugatti hue. An 'original' car with racing history.

Dashboard design was not ignored by Bugatti, the Scintilla magneto and the walnut wood wheel being of note.

Classic profile: Type 35 (with supercharger) seen in Bugatti blue with alloy wheels.

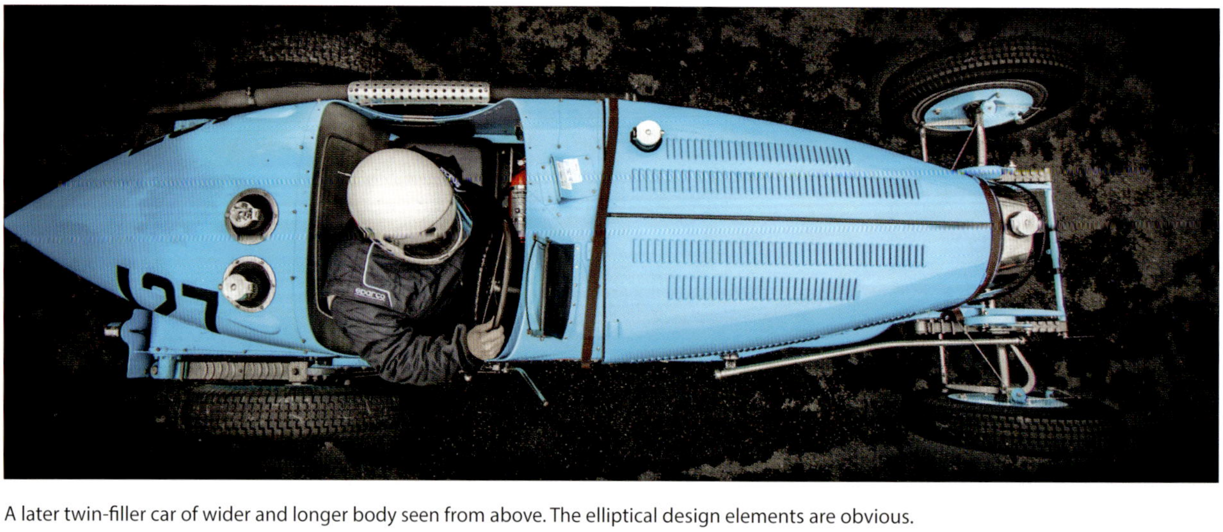

A later twin-filler car of wider and longer body seen from above. The elliptical design elements are obvious.

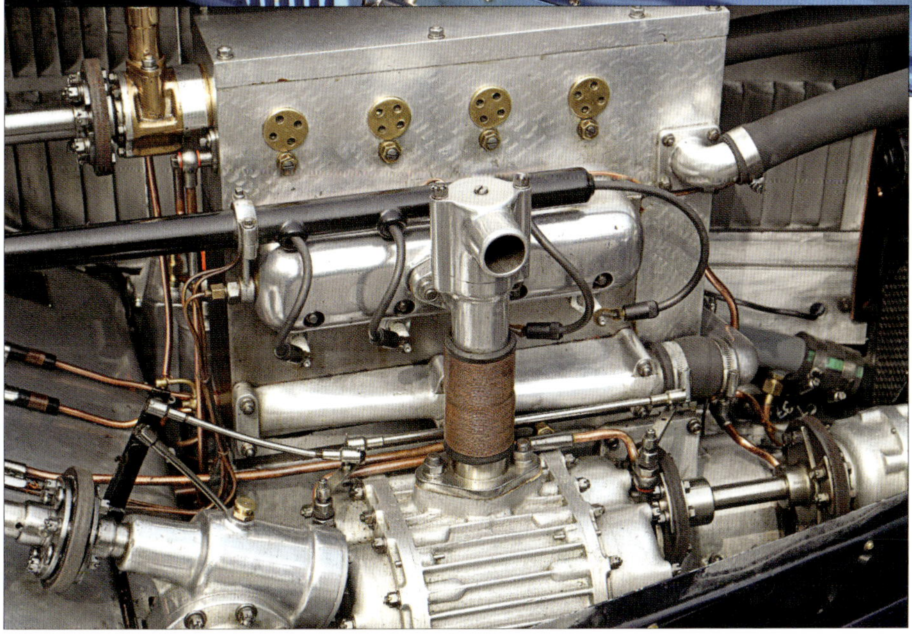

Top: Type 51 driven by Edmund Burgess going up Prescott hill with a front wheel raised, demonstrating the intent of the original chassis design.

Above: Supercharged, polished and a true example of Type 35 detailing.

Left: The smaller Type 37 four-cylinder engine with 'blower' provides under-bonnet contrast. Bugatti was known for his concern with external and internal details.

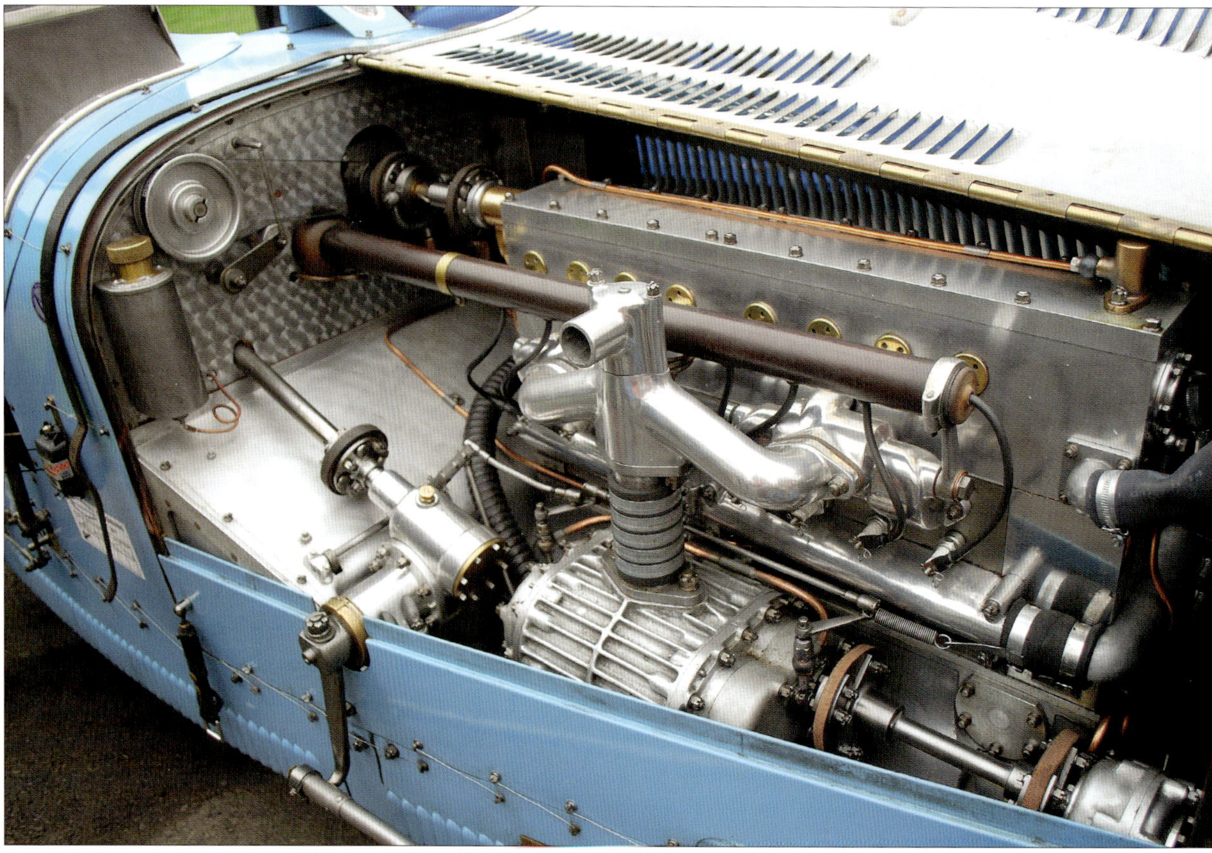

The later 4.9-litre engine as seen in the rare Type 54 as displayed by Ivan Dutton Ltd.

Above: Boat-tailed beauty in blue as she departs Prescott. The difference in rear-wheel-to-front-wheel track is clearly illustrated.

Left: Julian Eckersley at the wheel of a Bugatti at speed. Note the luggage bag carried in the correct manner over the spare wheel. The small scale of the car becomes clear in this view.

The essential Type 35 design details, beautifully and accurately captured in the work of internationally-known automotive artist Charles Duggan. Here the earlier 'narrow' radiator Type 35 also displays the sharp boat-tail, driver's external levers, the cast alloy wheels and correctly-aligned, body-mounted locking wires. The racing-specification Boyce-Moto-Meter temperature gauge is also fitted. The difficult-to-align front wheel angle has also been accurately achieved. The suspension details – leaf springs and clips – are expertly portrayed. Bugatti's use of leather in the design is also included in Duggan's rendering. An artistic tribute and a vital guide for modellers to follow.

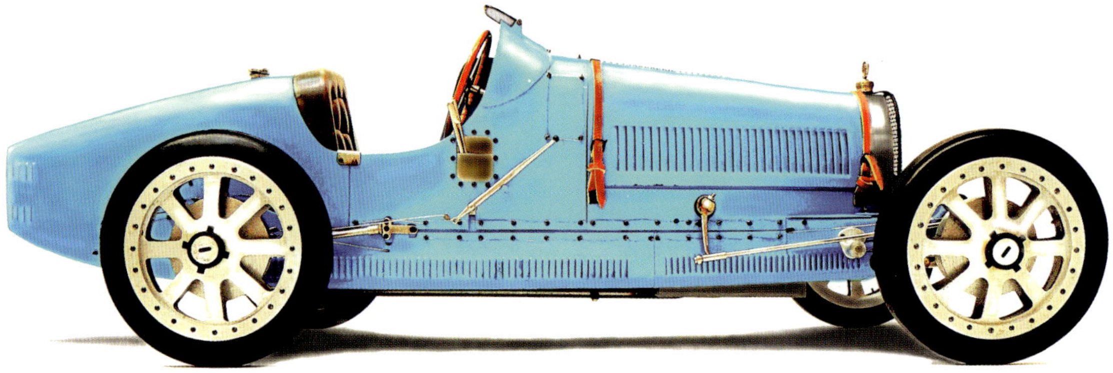

Ettore Bugatti made up his own paint hue formulations. His bright Bugatti blue was a different shade from French racing blue – it was lighter, brighter and had more green-to-turquoise in it. The 'Works' team cars used the blue, but Bugatti red was also popular as were other colours. Much earlier types of Bugattis were raced in white – reflecting the company's pre-1914 German collaborations (white being the German national racing colour). Some Type 35s were indeed raced in white or part-white including Hans Stuber's Type 35C. Here, in this Type 35 profile view, we see all the details of the design – body shape, steering arm rods, gear lever, louvres, vents, and an angled aero-screen. The beaded-edge variant wheels are 24-stud, indicating later-specification and larger brakes (earlier wheels were 32-stud).

Seen from above, the true elegance of the unique ellipsoid body shape is revealed. This view also includes the curved over-scuttle panel added to the passenger's side of the cockpit (to provide some form of protection and airflow improvement). A single-filler fuel cap fitting is evident in the rear body. The very fine gauges of the engineering elements can also be appreciated. Under the body, the chassis longerons are turned inwards. The passenger or 'riding mechanic' seat is thinner with less padding to reduce weight and to give more room. A Type 35 cockpit was very narrow and a tight fit for two. The non-driver would have to put an arm around the rear of the driver and hold on to the fuel filler for support.

The crucial details of the car are perhaps best captured in a front view. The 'bent' axle, leaf springs, brakes, steering-gear and shock absorbers are all designed to the closest degrees of tolerance and efficiency. Note the thin gauges of all components. Here at the front the chassis rails, or longerons, narrow and become so thin that they can flex – but are designed to do so. The cast alloy wheels with the inset brake drum design are also shown. The wheels are attach to the hub with the brake shoe integrally mounted. Note the inwards-taper of the body's nose profile and the brass and leather fittings that were so typical of Ettore's engineering-as-art skill set.

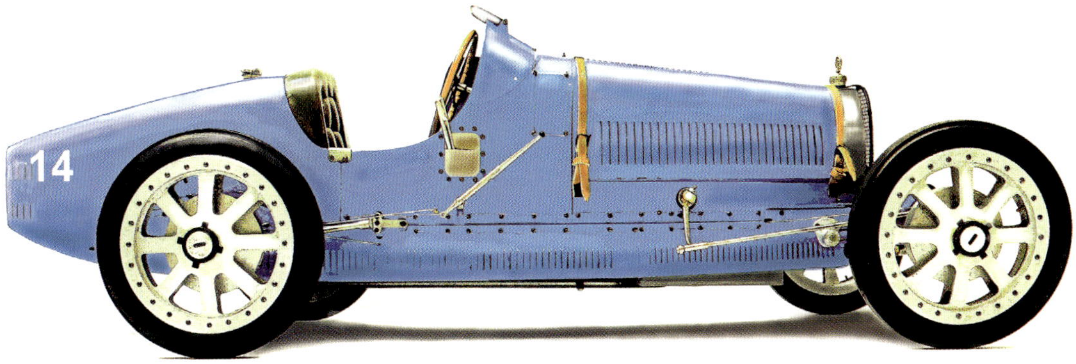

French racing blue shows its difference to Bugatti Blue in being slightly darker and bluer. This profile depicts a typical location for a stencil-applied racing number on the boat-tail, circa 1927. Some numbers were hand-applied in white wax paint. A more usual location for the racing number was the front body section of the car – where larger-sized numbers would be painted on or applied and often over the vented bodywork sections.

Above and below: The vibrant Italian 'racing' red was a later fashion. Earlier-period Fiats, Alfas and Maseratis often appeared in a bright blood red or a darker red hue. Bugatti created his own red palette – a brighter red and a darker hue that was not maroon yet not bright red. The darker red was used on several cars that were raced by British privateers such as Mays, Berthon, and Villers. British driver W. Grover-Williams raced a dark green Type 35. The famous Type 35C 'Genie' was finished in the brighter Bugatti red paint and retains its original hue to this day. In the darker red seen here, the number application represents a hand-applied technique using an easy-to-remove 'paint'.

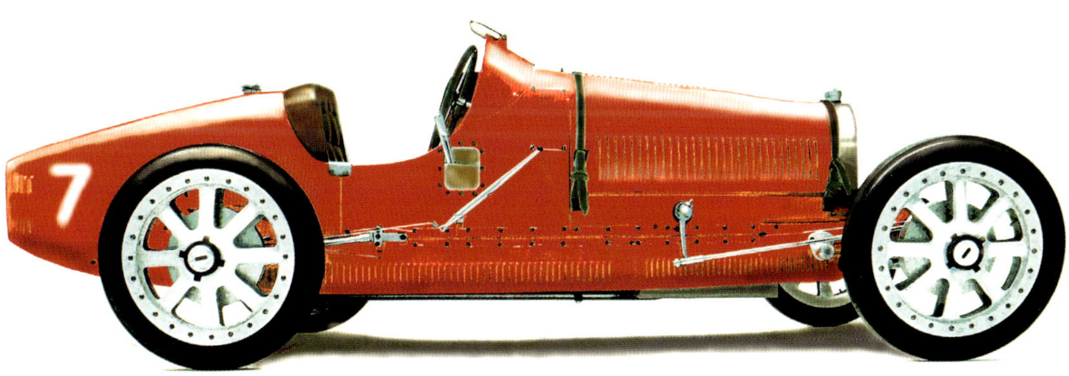

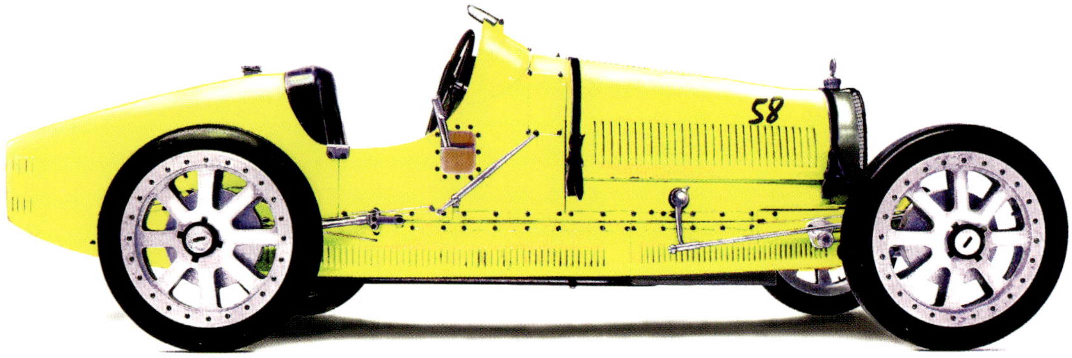

The Czech, Elizabeth Junek (Eliska Junkova), competed in her bright yellow 1928-purchased Type 35B and even led the 1928 Targa Florio until dropping back off the podium placings toward the end of the race. She used the number 58 and it was applied to the front of her race-specification car. She had earlier competed in a Bugatti Type 30 after marrying her Bugatti-racer husband Cenek Junek. For restorers or modellers, creating the correct hue of yellow is extremely difficult. The hue is not as bold as Giallo Fly of Ferrari type, neither is it like the modern Ford Daytona Yellow. Lemon Yellow is too pale. Reference to original colour photographs of the yellow Bugatti used in the 1920s is the best advice.

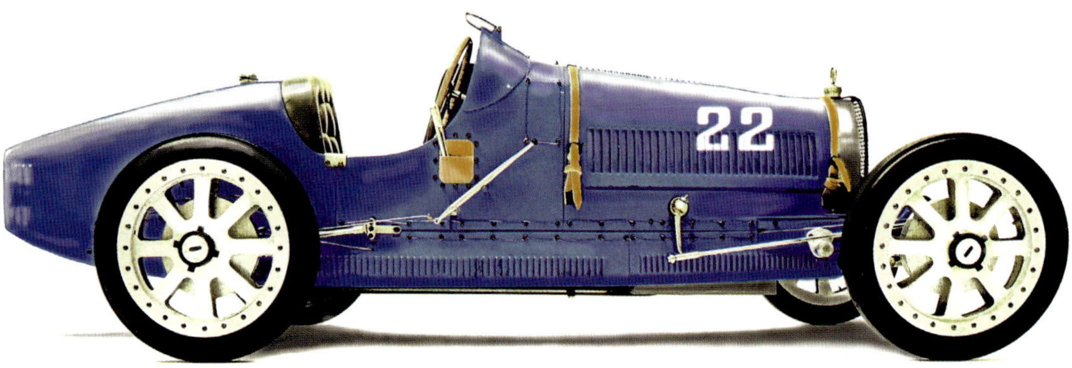

Type 35B with the number 22 was regularly used on a dark blue Bugatti, and was also used by Rene Dreyfus, including his win of the 1930 Monaco Grand Prix. In that race he fitted an extra fuel tank. Like so many race-specification cars, this one has a stone-guard fitted to the radiator. Various fonts were used for numbers and when applied to the front panels, the numbers over-wrote the bonnet vents as can be seen here.

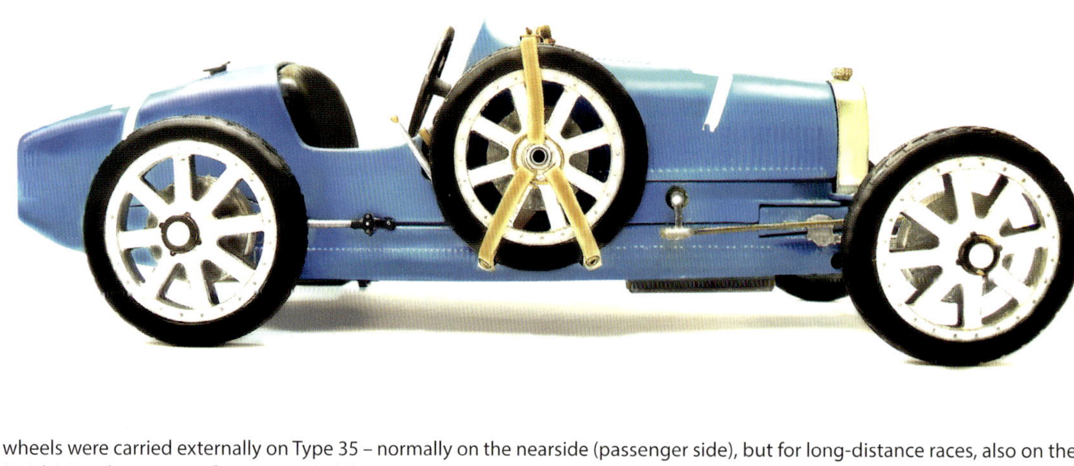

Spare wheels were carried externally on Type 35 – normally on the nearside (passenger side), but for long-distance races, also on the offside (driver's side). Leather straps, often in tan, dark brown, or more unusually black, were used to hold the wheel to a special bracket mounted on the bulkhead side panel. The offside-mounted spare wheel had to have a wooden wedge added to its mounting in order to splay it out away from the external gear lever.

AMALGAM COLLECTION (BRISTOL, UK) BUGATTI TYPE 35 (1926) 1/8 SCALE

Standing above all Type 35 models in the current marketplace is the Amalgam Collection brand (of Bristol UK) and its Type 35 at 1/8 scale. This wonderfully accurate and detailed model costs over £8,000, and so it should: it is modelling taken to a new dimension and unlikely to be surpassed. Collectors have flocked to it and as can be seen from the accompanying photographs, the realism is stunning. The details of the complex steering gear, body, and engine features all beyond 'model' expectations.

This resin/metal rendition is so close that one can almost hear it. It is tangibly a Bugatti in all but full-scale reality. The Amalgam Collection model is of Bartolomeo 'Meo' Costantini's victorious 1926 Type 35 – raced at the Targa Florio in 1926, where Bugatti dominated with a 1-2-3 victory. A close friend of Ettore Bugatti, Costantini retired from racing at the end of 1926 and took over as full-time racing team manager from Ettore. The Amalgam Type 35, resulting from digital scanning and studying the original Type 35, has been hand crafted and finished with the co-operation and assistance of Bugatti regarding original finishes, materials, archive imagery and drawings. The use of supremely accurate digital scanning of the original car allowed Amalgam to perfectly recreate every detail at scale. Detailed scrutiny by engineering and design teams ensured complete accuracy of representation. Is this the ultimate, defining Type 35 model? It has to be.

The Amalgam Collection's stunning model heads the international Type 35 model market. This model is of Bartolomeo 'Meo' Costantini's winning car as raced at the seventieth running of the Targa Florio in 1926, where Bugatti dominated. This model is as exquisite as the real thing. Expensive but probably the best Type 35 scale rendition yet seen.

The alloy body panels and tail can be seen here. Later Type 35s had a slightly fatter rear boat-tail shape and such design detail is a modellers' essential. This is the correct earlier rear body shape.

From the front, the sheer obsession with detail and correct scale rendition is obvious. Note the wheel and brake unit castings and the vents and louvres – all correctly presented. The locking wires also follow the correct factory pattern and are not diagonally woven.

This view reveals more of the cabin and dashboard details of the essential Bugatti design elements. Seating, dashboard, gauges, dials, as well as gearbox and chassis frame details, are all shown. Wonderfully scaled and just look at the seat detailing.

The Amalgam Collection's world-leading Type 35 model has been forensically modelled on the 'real thing' and offers a level of realism and accuracy previously unmatched. Note the superb suspension component detailing.

Engine detailing: seen from both sides, Bugatti's engines were 'square' rather than ellipsoid like his bodies. How close can a 'model' get? The answer is, via the Amalgam Collection, this close.

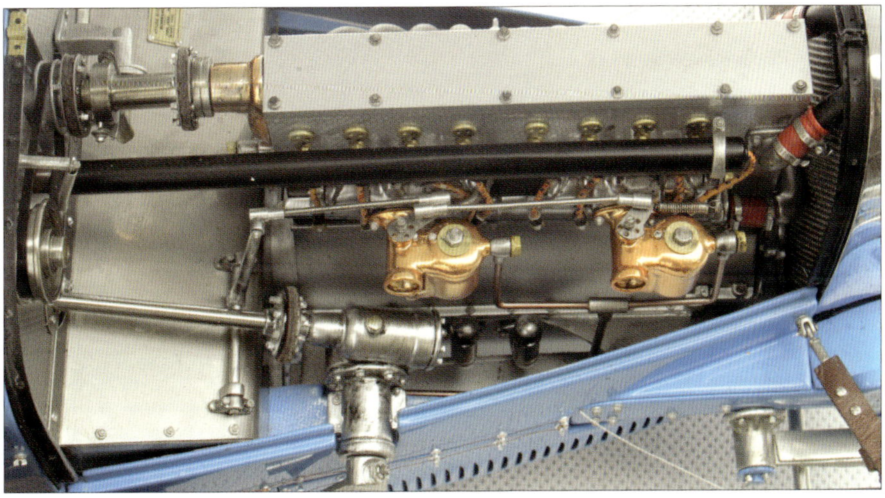

Under the Amalgam car's bonnet, the accuracy continues with the 'straight-eight'. Even the chassis identification plate is as per 'factory' spec.

In side profile view, the spare-wheel carrier on the driver's side is an unusual early racing detail. The piece of wood is used to tension the wheel in its straps away from fouling the external gear lever and handbrake. Note also the locking wires and the wheels. Digital scanning, forensic measurement and scaling all went into this 'work of art'.

The correctly-formed boat-tail and rear body details of the Amalgam Collection model.

Amalgam's superb representation of the cast alloy wheel, integral brake housing and suspension leaf spring and clips detailing

ART AUTO COLLECTION (J-P FONTENELLE, VICHY, FRANCE) BUGATTI TYPE 35 (1924) 1/8 SCALE

The Art Collection 1/8 scale Type 35 was also a strong player in the large scale field and now achieves serious money (in the thousands) on the open market. Known variously as the 'Art Auto Collection/Fontenelle Type 35' the model in metal was produced in small numbers (less than 50) and cost approximately £1,600. Of note, and seen here, is the curved, scuttle-cowl fairing. This was a detailed, advanced metal kit for modellers. The kit was hand formed, with bare brass body panels and correctly machined metal surfaces. Included were the alloy engine, chassis, and wheels. Cabin trim included leather seats, a wood-rimmed steering wheel and working steering, brakes and suspension. Dashboard and magneto details were well executed as was the walnut wood steering wheel.

The engine was accurate in scaling and included the correct fittings. Overall in this kit there were 466 parts.

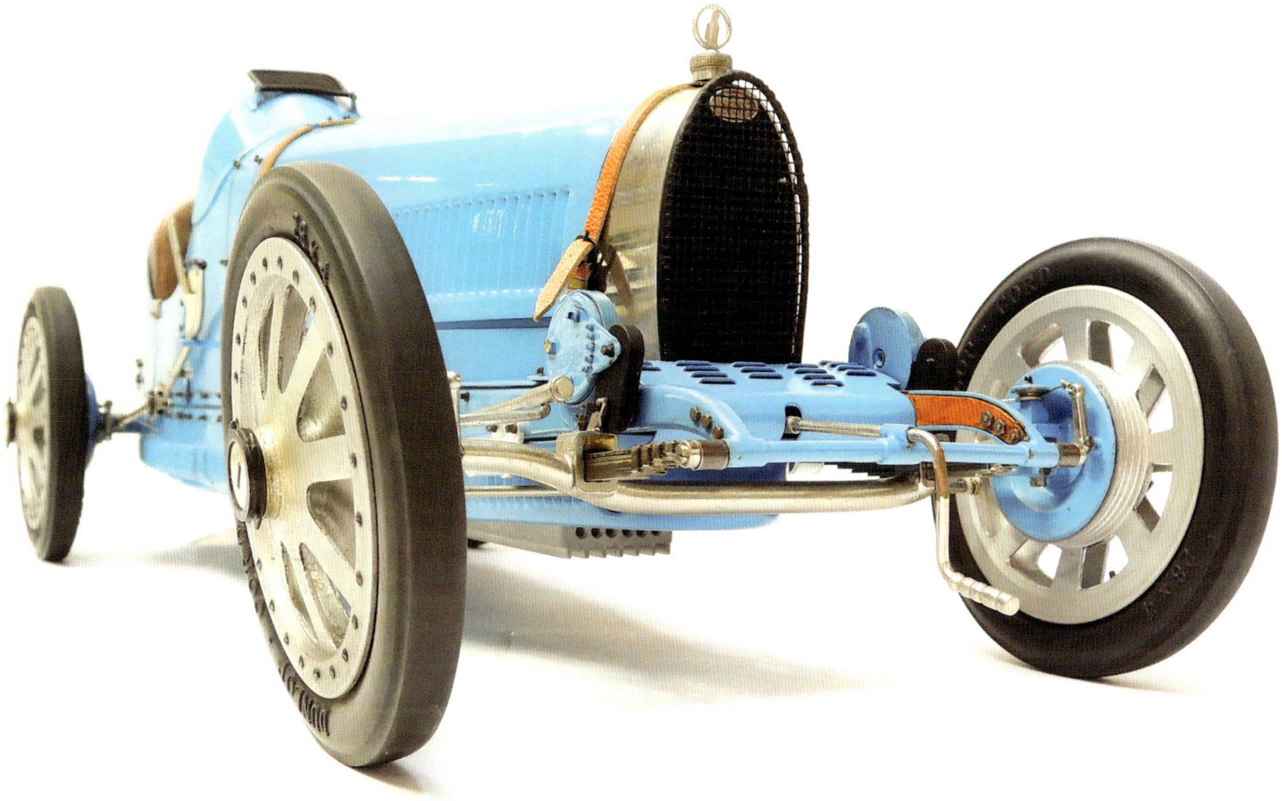

Art Collection's excellent rendition of the vital Type 35 elements is well captured from this angle.

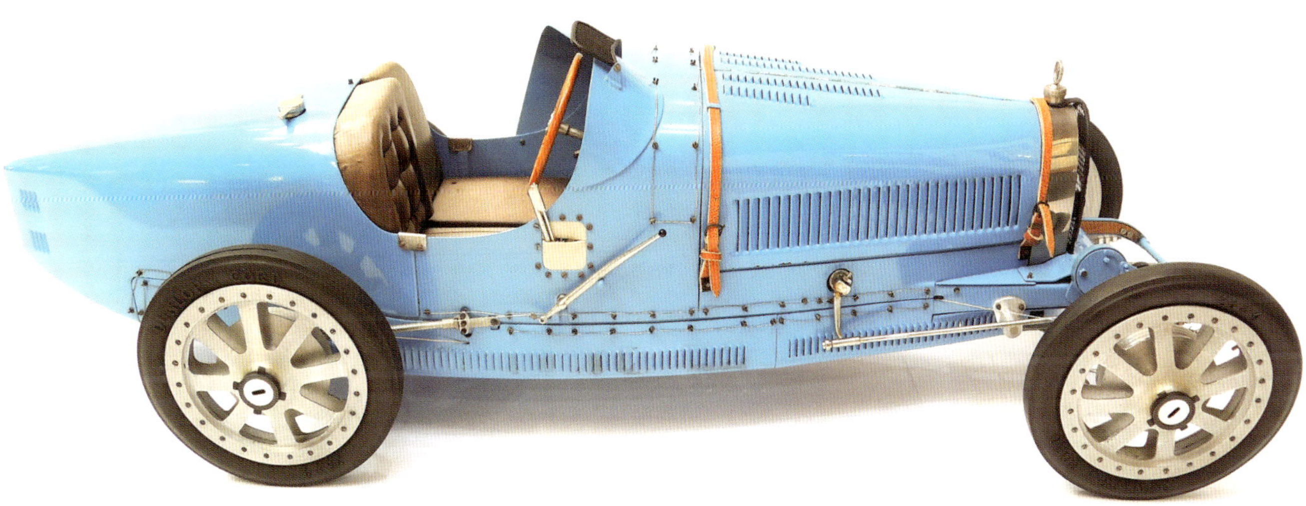

The Art Auto /Fontenelle 1/8 scale Type 35 in metal was produced in small numbers. This was a challenging advanced metal kit for forensic modellers. The kit was metal, with bare brass body panels and correctly machined metal surfaces. Note the use of the correct-pattern bolts and fasteners.

Art Collection's interior was impressive and included the rare, curved 'aerodynamic' extension panel covering for the secondary or riding mechanic's position.

Dashboard and magneto details were well executed.

The model featured the earlier, narrow radiator and 'short' bonnet. Steering and suspension were well detailed. This Art Auto kit, long out of production, now sells for big sums.

The driver's seat was padded for comfort, the riding mechanic's 'perch' was not – to save weight. Note the external gear lever and handbrake detailing.

The genius of Ettore Bugatti's design is easily grasped in this view of the Type 35. Every line 'works' and blends into an elliptical whole. Note how the nose and the tail both narrow – to reduce aerodynamic drag. This is the 1/8 scale Art Auto Collection model at its detailed best.

MONOGRAM (USA)
BUGATTI TYPE 35B
1/24 SCALE

Monogram's 1960s Type 35B model kit in plastic was remarkable for its price and quality even if the chrome work and some detailing were a little heavy-handed. It is still available on the second-hand-kit market. At the smaller 'standard' model scale, Monogram were off the mark early with a 1966 release of a Type 35B, at 1/24 scale. The launch edition box carried the name of the Bugatti Owners' Club Ltd as an official endorsement of the authenticity of the kit. Well-proportioned, reasonably moulded, and clearly the result of study, this kit (now re-tooled and re-boxed over many years, 1966, 1972, 1978, 1979, 1991), became an early benchmark. The car depicted was the 1930 Monaco-winning Type 35 of private entrant Rene Dreyfus.

However, there were some issues – notably the chrome-work effect which was coarse and overdone, and the definitions of the radiator and the instrument panel. The engine was pretty good though for a kit at this scale and just needed the correct painting technique to be applied. Modellers tended to create their own chrome effects, notably at the radiator, and to make improvements to the dashboard and instruments. Correct assembly and some tweaking would also get the front wheels set at the right angle – a factor that is often a fail point with Type 35 kits. Applying modern paints and decals can greatly improve the effect that can be created. Decal specialist VRM now produces decals to use on the 1/24 scale Monogram Bugatti T35.

Monogram's 1960s Type 35B model kit in plastic was remarkable for its price and quality even if the chrome work and detailing were a touch clumsy.

The correct Type 35B-spec details were part of Monogram's moulding. Modellers took to refining, filling, shaping and 'improving' some of the details – especially in later years when the kit's tooling perhaps showed its age in moulding quality. Work on steering, arms and suspension is often undertaken by modellers.

For a 1960s issue, the kit is well detailed. The real issue in terms of modelling forensics is the chrome work, radiator and eight-cylinder engine bay – but improvement is not difficult. Use of modern paints, decals and photo-etched products can greatly improve the standard model's chrome and metal work issues.

The boat-tailed rear and elliptical body seen from above, with the bonnet panels removed and the engine exposed.

Monogram's model finished to a high degree. The chrome and alloy treatment could be further improved by deploying the very latest photo-etched materials.

Monogram's 1960s Type 35 model kit allows the modeller to create a good Type 35 which can now be improved upon using more modern parts and accessories.

REVIVAL (ITALY)
BUGATTI TYPE 35B
1/20 SCALE

Revival (Italy) produced an excellent, die cast pre-painted metal 1924–1926 Type 35A at 1/20 scale that, with minor modifications, could also be made to represent a Type 37. Targa Florio, Tecla and the standard Type 35 could all be created. Kit numbers are: REV 80101, 80102, 80103. Of interest, the kit could be built with the two-spare tyres set-up for long distance road-racing.

The Dunlop tyres are superbly rendered. Perfectly scaled alloys graced the Type 35 launch kit, while the Type 37A (depicted in yellow paintwork) featured Revival's version of the fine-spoked wire wheels. The Revival Type 35 kit came as a new tooling then with the two new box issues with new variants and parts. Unused, perfect boxed kits are currently on sale at £300–£500.

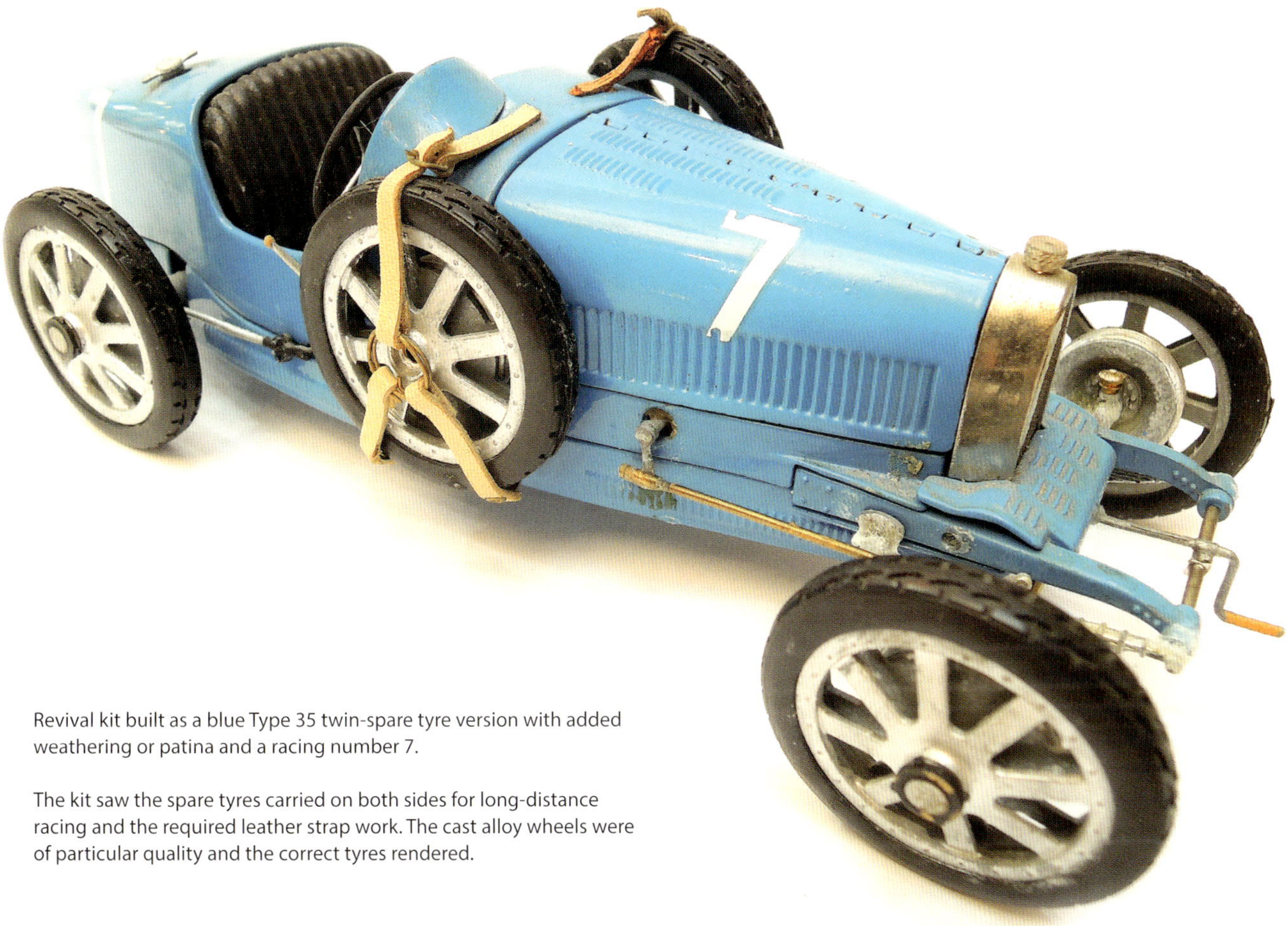

Revival kit built as a blue Type 35 twin-spare tyre version with added weathering or patina and a racing number 7.

The kit saw the spare tyres carried on both sides for long-distance racing and the required leather strap work. The cast alloy wheels were of particular quality and the correct tyres rendered.

The brake drums might be slightly over-sized in casting terms: suspension details were well executed.

Revival produced a Type 35 with black rather than tan seat trim and a non-standard black-rimmed steering wheel. The black leather seats are more unusual than the tan leather so often seen on blue Bugattis.

The classic view from above shows off Revival's capturing of Type 35.

The detailing of Ettore's advanced alloy wheel design is well defined in this model.

The earlier Type 35 used the narrower radiator and shorter front body section. Later cars saw changes that moved the radiator forwards and made it larger – reducing the size of the front apron area.

Modelling the Bugatti

Universally loved, Type 35 has been a modeller's favourite for decades. There are several websites for Bugatti enthusiasts and Herman Brouwer's superb bugattibuilder.com has to be cited as a point of expertise and debate for the Type 35 model builder. Herman's expertise is well known. Other modelling forums exist and the website bugattipage.com includes modelling forum discussions. Scalemodels.co.uk and modelingmadness.com also provide web-based Bugatti modelling discussions.

In the era of resin, and die cast models, the 1960s to 1970s are believed to be the starting point for commercially produced models of the Type 35. However, the truth is that the first scale model of a Bugatti Type 35 was built by no less a person that Ettore Bugatti himself. In the late 1920s he built a wonderfully accurate Type 35 pedal-type, electrically powered ride-on model car and young Roland Bugatti can be seen in photographs using this perfectly rendered scale Bugatti. In fact, Bugatti produced several hundred of these electric 'toy' cars.

More recently, the Schlumpf Collection of Bugattis did of course provide many model manufacturers and many model makers with inspiration and although now structured in a new museum context, it remains a vital reference point.

As early as 1930, the Paya Company of Spain mass manufactured a tin-plate 'toy' Bugatti Type 35. To avoid concurrent rights issues it was labelled 'Bugat'. Founded in 1902 by Rafael Paya, Rafael's sons built the first toy factory in Spain and by the 1920s Paya's toys were considered of excellent standing across Europe and rivalled German models. Paya's toys' liveries were of note. In the 1970s a limited re-launch was issued. Reproductions now stem from China.

In the 1960s, Pyro of America produced an early Type 35-based Grand Prix kit at 1/32 scale and it was an early, if basic, moulding.

Private modellers also created expensive, hand-built, one-off metal Type 35 models, tin-plate versions and high quality, large, 1/8 scale Type 35s. Gerald Wingrove's Bugatti models were beyond exquisite and one is now exhibited at the Bugatti Trust.

So popular is the Type 35 across all car marque enthusiasts, that today it has also spawned incredibly accurate and expensive, large and mid-scale models that are as much works of art as the real car.

In the 1980s, the Franklin Mint produced a die cast 1/24 1924 Bugatti Type 35. This was well detailed with an excellent engine bay and a nickel-plated radiator (as per the original car) and well-executed cabin castings and floor members under a removable seat. A small scale but respected model.

More currently, Romania's Wespe Models 1/8 Special Big Scale-branded Type 35 in resin, is an excellent access point for the Bugatti modeller at larger scale, but is not cheap. Wespe's model remains an accessible route to enthusiastic Bugatti modelling.

Grand Prix Models in the UK market a 1/43 scale Bugatti Type 35 Tecla 6-Wheel Hillclimb car that is nicely specific. The depiction of Bugatti's rare use of double rear wheels for hill-climb traction is noteworthy for the Bugattiste.

South Eastern Finecast, Wills, and SMTS have all issued metal Type 35 models kits.

Germany's CMC Models (below) have recently launched a superb range of Type 35 metal-bodied renditions, notably in a series of national racing colours. The detailing, scaling and colours are all superb and at the price of several hundred Euros per

car (approximately £350), provide the committed Type 35 enthusiast with a strong anchor for their interest. This is serious metal kit building for the enthusiast with over 500 parts per kit and some great livery options.

Die cast and white metal Type 35 kits have provided modellers with a basis of build and modification over the years and Scale Model Technical Series issued a metal kit that is still available second-hand. Burrago's somewhat heavy-handed rendition in metal is not to be dismissed because with work and refinement of certain features, it can be turned into a quality display item. Eligor (Type 35C 1928), Platignum and Soligor are Type 35 kits that are also well known and are widely traded on the kit market.

Burrago manufactured a low-cost metal Type 59 which was part of their Bugatti series. Again, it is not a Type 35 but it is a descendent of the boat-tailed series that began it all. Not quite a 'collector's item', but a Bugatti model none the less.

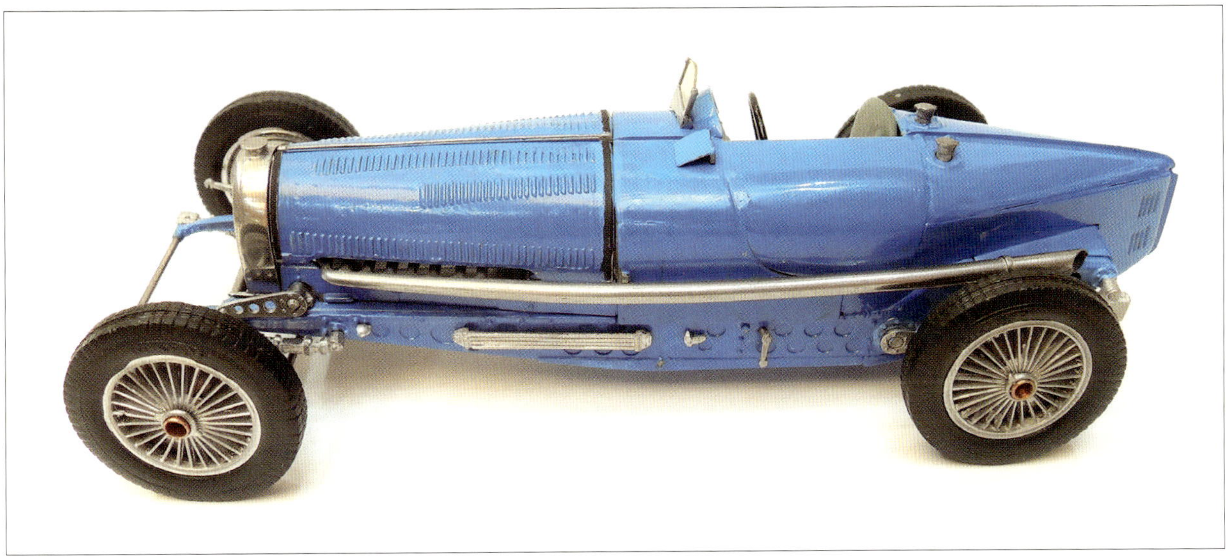

Type 59 was a different beast yet continued the original ethos of Type 35. This Brumm model captured the essence of the beast. The wheels and rear 'spine' are of note.

But we must look to Monogram and to Airfix for the first true, volume production mass market plastic Type 35 model kits at 1/28 and 1/32 scales in the modern era. Cheap, and perhaps deemed as basic by some, they offer access to the modeller whose means do not extend to hundreds or thousands of Euros, Dollars, Pounds or whatever their local currency, for a single model. The Airfix and Monogram kits also allow plenty of personal refinement and dare we say it, customization. Revell's Bugatti kits do likewise.

Airfix first launched their Type 35 kit in 1975 and this had been reboxed, rebranded, and fettled over the years up to 2004. The body proportions were good but the vents and louvres rather heavy. Wheel detail was of note, but the expert modeller will of course refine and hone numerous aspects of the moulding and features. As with the Monogram kit, the aero-screen gauge was thick but could easily be replaced by a more accurate gauge of 'glass' representation. Both road and racing specification versions of the Type 35B could be built from this kit and a driver figure in period attire was included.

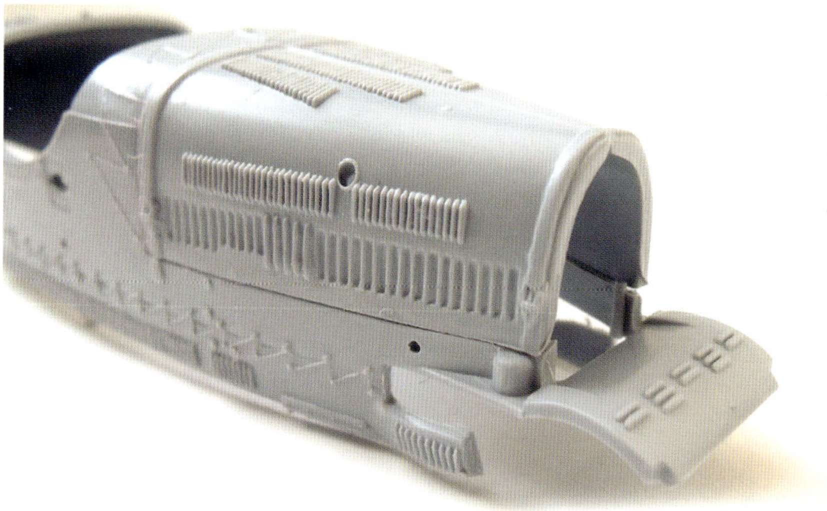

The Airfix body was accurately scaled but required work on refining mouldings for vents and component mountings. At this scale, refining the louvres and vents is not an easy task.

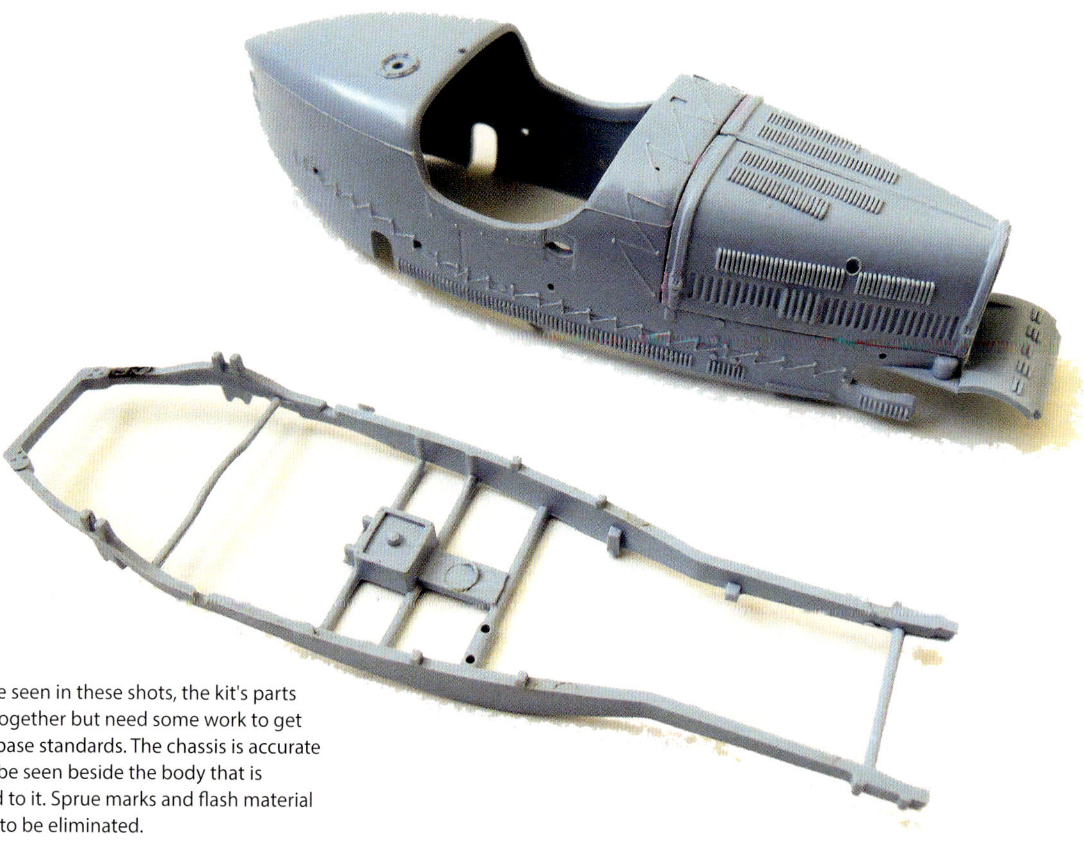

As can be seen in these shots, the kit's parts go well together but need some work to get beyond base standards. The chassis is accurate and can be seen beside the body that is mounted to it. Sprue marks and flash material do need to be eliminated.

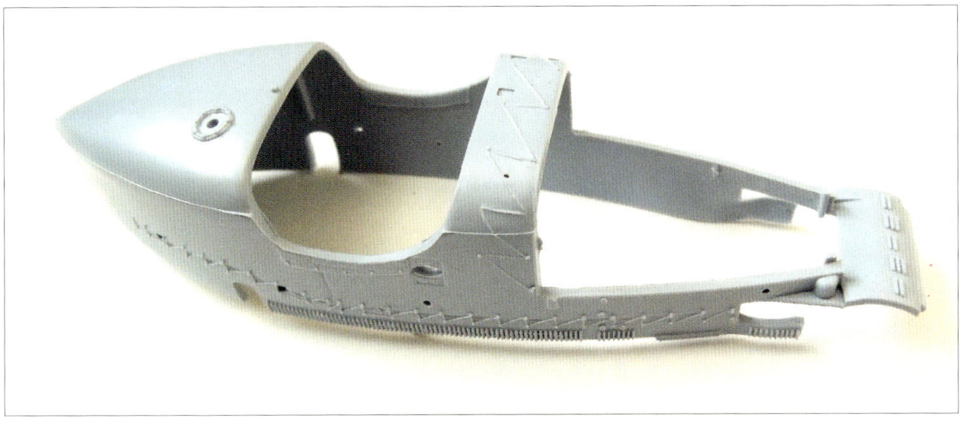

Airfix produced its Type 35B at 1/32 scale from 1975. The body and chassis details were well represented but suffered from slightly coarse detailing to the vents and panel shuts – which the modeller can refine. A road-going version with lamps and mudguards/wings was part of the kit – as was a driver.

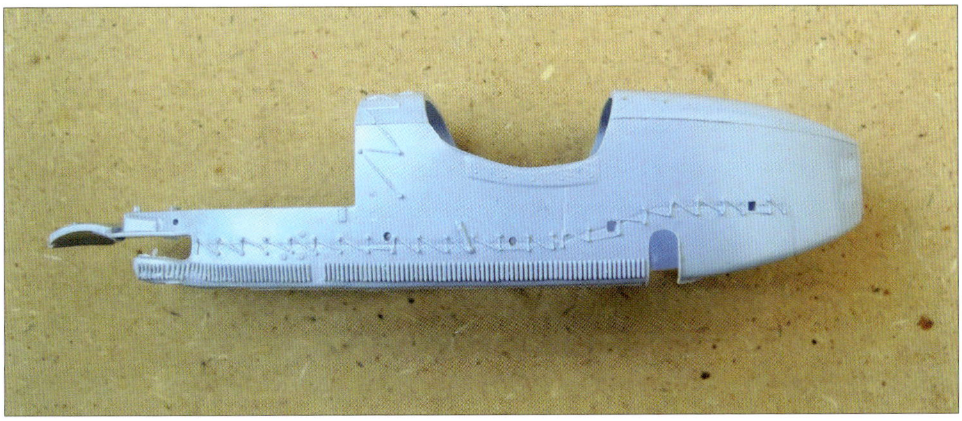

Airfix Type 35 body featured the alternative diagonal-pattern locking wiring. Moulded-in, the forensic option is to paint each 'wire' silver – no easy task at this scale.

Side and rear view of the DB Models Type 35-based, special-bodied car with bulbous styling and extra panels. A scarce model of very rare and much-discussed, USA-origin external variant upon the Bugatti Type 35.

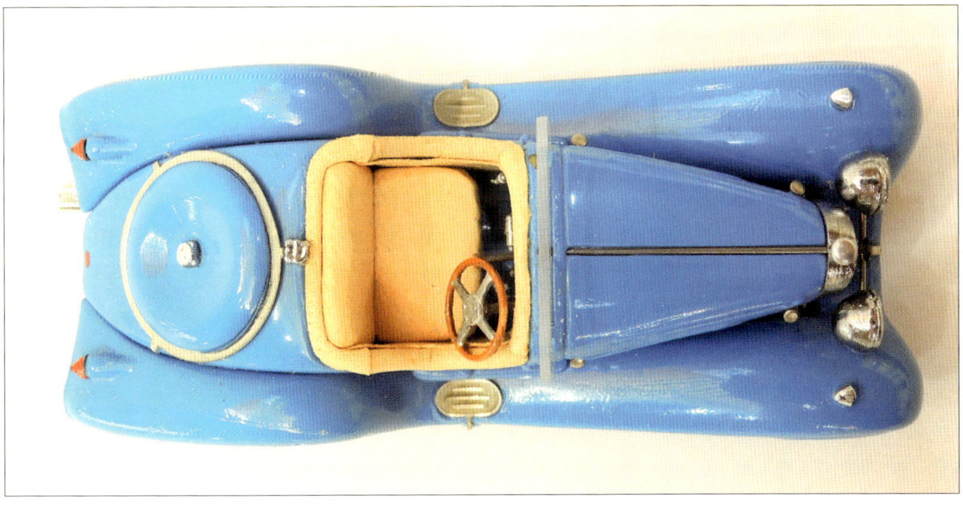

See from the top, DB Models' special-bodied roadster edition.

See from the ront, DB Models' special-bodied roadster edition.

DB Models have produced Type 35s and a special-bodied roadster rendition. Revival produced a Type 35 Tecla variant and other derivatives at 1/20. Revell also manufactured at Type 35, but no current Tamiya Type 35 is evident. At the other end of the scale, Harbutt's Plasticine Company marketed a moulded Type 35 in this material.

Bandai (marketed as Bugatti Model 1932 Type 35 Super Sport), and Italeri, are both brands that have produced Type 35 models over the years. Many more across all scales exist including metal, die cast and plastic/resin, and are listed below.

VRM decals and Renaissance provide Bugatti modelling accessories and extras.

Type 35 models

Airfix	Matchbox
Amalgam	MCM
Ami	Monogram
Antonetti	Paya
Art Collection Auto	Piccollino
Bachmann	Platignum
Bandai	Pyro
Bouisso	Renaissance
Bouwe de Boer	Revell
Burrago	Revival
CMC	Scale Model Technical Series
Corgi	Solido
DB	South Eastern Finecast
Eligor	Ulrich (Bronzes)
Franklin Mint	Vroom
Grand Prix Models	Wespe
IDEM	Wills
Italeri	Wingrove
John Day	Renaissance
Lesney	

Colours

Ettore Bugatti mixed his own colours to his own chemical formulas. His famous house/team colour of blue was not French Racing Blue nor a turquoise tinted sky blue. Photographs of cars can be misleading in colour due to the effects of strong sunlight causing original-era and modern cars to fade, and there are photographic limitations. Modellers can consult web forums (such as bugattibuilder.com) to access debate and information about mixing the correct blue hue. It should be noted that many modellers use a slightly darker shade of blue that is close to the Garter Blue used on the not-unrelated Gresley A4 Pacific Class LNER 1930s locomotives.

Bugatti Blue requires a careful mixing and correlation. A Tamiya blue can provide a good base hue for modification. A darker blue shade more akin to Porsche's 1970s Arrow blue is often applied (perhaps erroneously), or a more green-hued, lighter turquoise shade is often used. A brighter shade of French Racing Blue can be used. ModelMaster French Blue has been cited but must require re-blending to add the lighter, brighter, very slightly green spectrum hue to create true Bugatti Blue.

Of interest are the other hues seen on Type 35s in-period and now rendered via modelling. Racing red, German Racing White, Dutch orange, the Junek bright yellow, black, dark green and maroon all feature across the historical palette. The very unusual mid-green (almost a duck egg plus moss green mix) of the British racing team named Scuderia Lemon Burton would make a rare Type 35 paint scheme. And, CMC models recent National Colours Type treatment has been very popular.

The expert modeller will of course closely research their chosen colour scheme across more than one reference source.

Sanding, filing and filling are of course basic skills to be deployed. The stripping of fake chrome and reapplying using a 'metalliser' of flat aluminium, and chrome silver can add realism. Adding sheen to leather effect paint for the seats using a polishing oil is often done. With the Type 35's delicate and intricate steering and braking gear and cables, modellers can choose to substitute more precision than standard mouldings apply. Fine wire is the choice. The issue of applying weathering and patina is an entirely personal choice but can create the more individual model – as can setting the model in a display or diorama. Applying patina is an individual issue for modellers, and seeing a Type 35 'dirty' as if having competed in a hot and dusty race, surely adds to the art of modelling. Numerous techniques exist for the modeller to apply to 'raced' Type 35s. Applying wear to tyres, cables, paint, seats, alloy or wire wheels and chrome work, amid application of mud and dust, offers great choice for the modeller.

Type 35-specific modelling-build and finishing discussions can be found across numerous modelling web forums and bugattibuilder.com is a great starting point.

As can be seen in the accompanying photographs, Type 35 is a modeller's delight and offers many variants to model with respective accuracy of specification to be achieved.

Be it a model, or the full-scale car, Type 35 remains a leading icon of car design and motor sport. It is deservedly revered and popular beyond enthusiasts' individual marque preferences. Type 35 stands tall above so many cars and their design and histories. Even Ettore Bugatti himself might have been surprised.

Key Type 35 model maker's issues:

Front wheel angle: use photos of real cars to assess correct angle.
Chassis: refine the rails and paint in correct Bugatti colours.
Louvres and vents: re-work and hone by hand.
Aero-screens: replace with thinner section.
Brake cables – use ultra-fine wire or, on larger scale models, guitar wire or metal fishing line.
Chrome work excess: replace with thinner section or etched.
Dashboard: improve fascia and instruments by hand or decals or etched. Check magneto location relevant to Type 35 specification.
Seating: correct colours and textures to leather or vinyl.
Steering wheel: requires correct walnut rendition in paint, not orange cherry wood shade.
Starter handle: refine from over-scale moulding.
Bonnet: refine louvres. Fit leather straps.
Tyres and wheels: ensure correct wheel and tyres types and sizes as appropriate to model. Check brake drum size. Note beaded edge racing tyres and the extra studs on some alloys. Note over-scale tyre treads supplied by some kits. Spare tyres x1 or x2 carried on leather straps.
Front axle: refine and rework as necessary. Ensure leaf spring clips are accurate.
Body work and undertray: hone, file and fill where appropriate.
Seating: options and riding mechanic to be considered.
Radiator, mesh stone protector: create in wire.
Exhaust: modify and site as appropriate.
Road gear: fit or delete wings/fenders and lights.
Moulding dimples and sprues. Refine and work as necessary.

The diversity of Bugatti modelling is captured by this view depicting the Brumm Type 59 Grand Prix racer (left) and the DB Type 35 special body touring car (right). Both are smaller scale, rarer items on the second-hand model market.

The Type 59 model (manufactured by Brumm) shows the longer, lower design and build of Jean Bugatti's later Grand Prix car. Note how the rear suspension is externally mounted – rather than internally enclosed on the Type 35 series. The 'spine' is also evident.

DB produced this 1/43 scale interpretation of the unusual, externally produced, modified Type 35 special touring body seen on full-sized cars circa 1927. Today it is a rare item that is one for the Bugatti purist.

Prior to his Type 35, Bugatti produced his Tank racing cars with pontoon bodies in the early 1920s. In the 1930s, a few Type 57G bodies of Tank design provenance were launched. This 1/18 scale T57G Le Mans model manufactured by Spark, celebrated the drivers Wimille and Veyron, driving the car a Le Mans in 1937.

Acknowledgements/References:
The author (a member of the Bugatti Owners' Club and The Bugatti Trust), would like to thank the following sources for vital information, reference material and kind assistance with content, photography, Bugatti-access and research: Bugatti Owners' Club, Charles Trevelyan (chairman), Chris Rogers (general manager), Paul Gibbons (archivist), Fiona Rogers. Similarly to The Bugatti Trust, Angela Hucke (curator), David Morys (archivist), Sue Shaw (secretary). Thanks also for information and access to: Brooklands Museum, Edmund Burgess, Stephen Gentry, Tim Dutton, Adam Ward, Mike and Sue Marshall, Chris Jacques, Julian Mazjub. Bibliographic and reference sources include: *Bugatti Blue: Prescott and the Spirit of Bugatti* by Lance Cole (Pen and Sword Books). *Bugatti: the Man and the Marque* by Jonathan Wood (Crowood Press). *Bugatti: Great Marques* by H. G Conway (Octopus Books). *Bugatti Type 35 1924 onwards /Manual* by Chas Parker (Haynes). *The Amazing Bugattis* (and contributors) The Design Council. *Bugantics. The Type 35 Grand Prix*, by Godfrey Eaton (Profile Publications). Main model and modern-era Bugatti car photography by the author. Others, BOC and Bugatti Trust archives. Illustrations by LFC Design. Editorial and design layout by Kerrin Cocks. CMC Models supplied the Type 35 national colours photos. Thanks to bugattibuilder.com. Many thanks to Amalgam Collection of Bristol, UK, for supply of 1/8 scale model studio photographs.